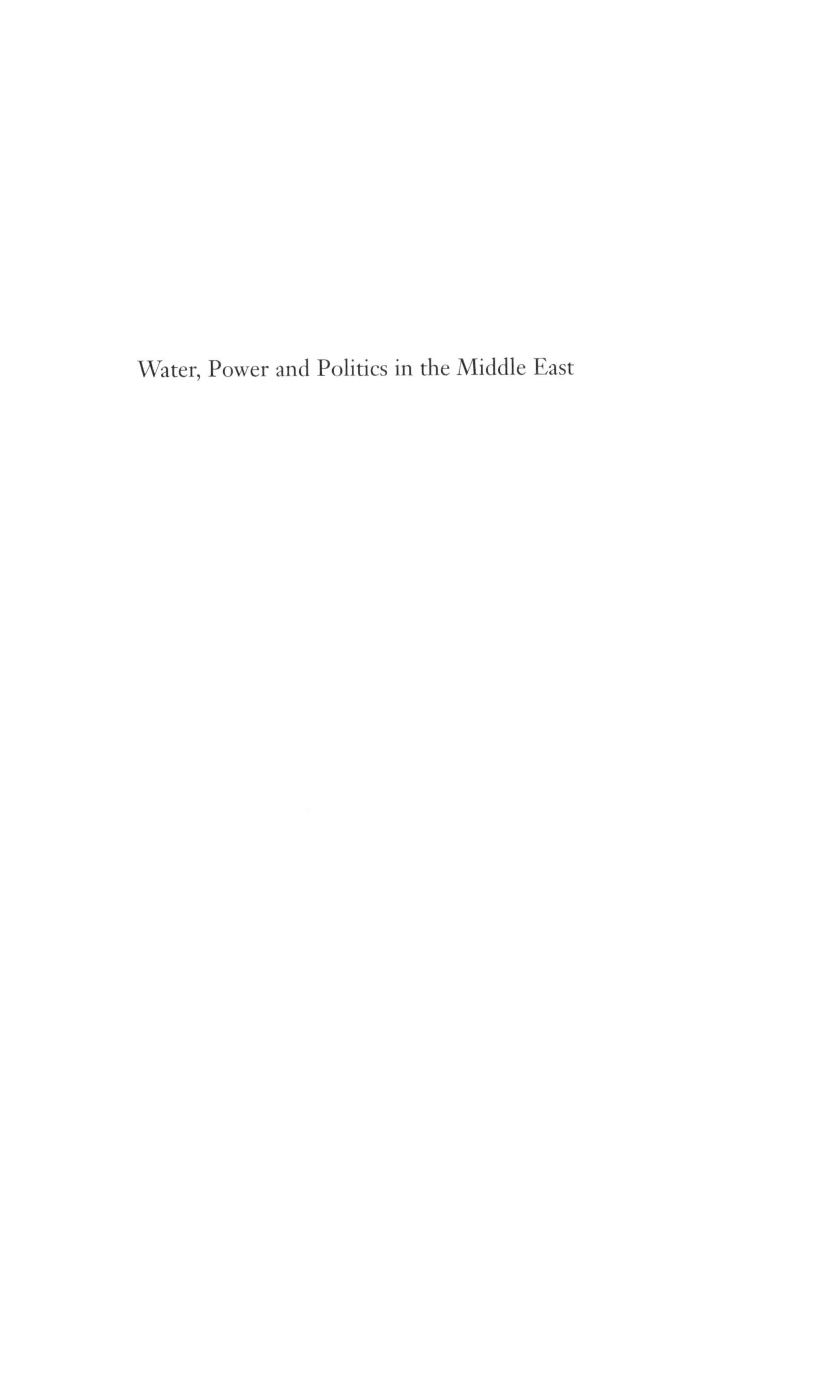

Water, Power and Politics in the Middle East

Water, Power and Politics in the Middle East

The Other Israeli–Palestinian Conflict

Jan Selby

Published in 2003 by I.B.Tauris & Co Ltd
6 Salem Road, London W2 4BU
175 Fifth Avenue, New York NY 10010
www.ibtauris.com

In the United States and Canada distributed by Palgrave Macmillan, a division of St. Martin's Press
175 Fifth Avenue, New York NY 10010

ISBN 1 86064 934 3

A full CIP record for this book is available from the British Library
A full CIP record for this book is available from the Library of Congress
Library of Congress catalog card: available

Typeset in Janson by Dexter Haven Associates Ltd, London
Printed and bound in Great Britain by MPG Books, Bodmin

CONTENTS

FIGURES

Acknowledgements

I was first introduced to the issues with which this book is concerned in August 1994. Having spent the early summer trying to teach English to Palestinian 13-year-olds in UN schools in Amman, I had decided to search for some less taxing kind of work in the West Bank. The Oslo process was up and running, and I felt sure I would be able to find work within one or other of the countless Non-Governmental Organisations operating within the Territories. I spent a week or so ringing around and knocking on doors, mostly in vain, until one afternoon I turned up at the Applied Research Institute of Jerusalem (ARIJ) to meet its Director, Jad Isaac. I left not only with a job, but also with a pile of papers and project documents on Israeli–Palestinian water issues, and would come to spend four months there as a researcher, writer and sometime proof-reader. It was during this time that I got to know the southern West Bank and its water problems, the main focus of this book. Little did I think, though, that eight years later I would still be writing on the subject.

This book itself started life as a PhD. Elizabeth Shove and Peter Wilkin were fantastic supervisors, both sharing a strong practical sense and a keen eye for overstatement. Thanks in particular should go to Elizabeth for prompting me to think about the sociology and politics of technologies and everyday practice, and to Peter for guiding me into the field of international relations. Thanks also to Mick Dillon, Mike Michael, Simon Shackley and Brian Wynne, all of whom provided supervisory help during the early stages of my research; to Gerd Nonneman, always a continual fountain of knowledge and enthusiasm; and to my PhD examiners Christopher Clapham, Tim Niblock and Julian Saurin, who since then have continued to give me invaluable and much-appreciated support.

The initial research for this book was funded by the UK Economic and Social Research Council, and I have also benefited from grants from the Centre for Science Studies, Lancaster University, to cover a short period in Israel and the West Bank during August 1999, and from the Faculty of Social Sciences, Lancaster University, to cover an equally brief research trip during June 2000. Thanks to each of them, and more broadly to Lancaster University for its institutional support during my seven years studying and working there.

In Israel and the West Bank, special thanks go of course to Jad, as well as to Yasser Abed Alqfar, Karen Assaf, Ihab Bargouthi, Ibrahim Dajani, Musa Al-Khaldi, Walid Kawasmi, Hilda Khamis, Tigritte Laham, Samira Rafa'ai, and David Scarpa. Thanks also to all of my interviewees, as well as to ARIJ, IPCRI, the Truman Institute and the Palestinian Water Authority for permission to use their libraries; and in London to Tony Allan of SOAS, Greg Shapland of the FCO, and Chris Saunders of Save the Children. Thanks to Ian Gulley at Aberystwyth for his help with the maps, and to David Stonestreet of I.B.Tauris for being willing to take up this project. Needless to say, none of the above bear any responsibility for any follies, indiscretions or errors of interpretation that might be contained herein.

The bulk of chapter 4 was first published as 'Dressing up domination as "cooperation": the case of Israeli–Palestinian water relations', in *Review of International Studies*, vol. 29, no 1 (January 2003), pp. 21–38. Thanks to Cambridge University Press for permission to reprint this material.

Finally, thanks to David, Bärbel and Rowena for their love and affection; to Steph and Mark for the Timble crack; to all my football, anti-war and university friends in Lancaster, who are greatly missed; and to new friends and colleagues in Aberystwyth for having helped to make my move to Wales such a painless, even happy, experience.

Abbreviations

ANERA	American Near East Refugee Aid
ARIJ	Applied Research Institute of Jerusalem
bcm/bcmy	billion cubic metres/billion cubic metres per year
CDM	Camp Dresser and McKee Inc.
CIA	Central Intelligence Agency
cmy	cubic metres per year
DfID	Department for International Development, UK
DFLP	Democratic Front for the Liberation of Palestine
DoP	Declaration of Principles
EAP	Emergency Assistance Program
FCO	Foreign and Commonwealth Office, UK
GIS	Geographical Information Systems
GTZ	German Agency for Technical Development
ICRC	International Committee of the Red Cross
IDF	Israeli Defence Forces
IDRC	International Development Research Council, Canada
IHS	Israeli Hydrological Service
IMF	International Monetary Fund
IPCRI	Israel/Palestine Center for Research and Information
IR	International Relations
JMCC	Jerusalem Media and Communication Centre, Jerusalem
JSETs	Joint Supervision and Enforcement Teams
JWC	Joint Water Committee
KfW	Kreditanstalt für Wiederaufbau
LEKA	Lyonnaise des Eaux Katib and Alami
mcm/mcmy	million cubic metres/million cubic metres per year
MOPIC	Ministry of Planning and International Co-operation, PA
NGO	Non-Governmental Organisation
NIS	New Israeli Shekel
NRA	Natural Resources Authority, Jordan
PA	Palestinian Authority
PARC	Palestinian Agricultural Relief Committees
PASSIA	Palestinian Academic Society for the Study of International Affairs

PCC	Palestine Central Council
PECDAR	Palestinian Economic Council for Development and Reconstruction
PFLP	Popular Front for the Liberation of Palestine
PHG	Palestinian Hydrology Group
PLC	Palestinian Legislative Council
PLO	Palestinian Liberation Organisation
PNC	Palestine National Council
PWA	Palestinian Water Authority
SOAS	School of Oriental and African Studies, London
UNDP	United Nations Development Programme
UNRWA	United Nations Relief and Works Agency
USAID	United States Agency for International Development
WBWD	West Bank Water Department
WEDO	Water and Environmental Development Organization, Bethlehem
WESC	Water and Environmental Studies Centre, An-Najar University, Nablus
WRAP	Water Resources Action Programme
WSERU	Water and Soil Environmental Research Unit, Bethlehem University
WSSA	Water Supply and Sewerage Authority, Bethlehem
WZO	World Zionist Organization

Introduction

OSLO: DEAD IN THE WATER

On 29 March 2002, Israel launched its Operation Defensive Shield, effectively bringing the nine-year-old Oslo 'peace process' to a close. Within a couple of days, tanks had rolled into all but a couple of the major West Bank towns. Yasser Arafat's compound in Ramallah was being besieged. Ramallah, Jenin, Nablus and other West Bank towns had been declared closed military areas, subject to strict curfews. Checkpoints had been tightened, and movement between Palestinian towns had been brought to an almost complete halt. Inside the sealed-off towns, the Israeli Defence Forces (IDF) were going about their business, their stated aim being to destroy the Palestinian 'infrastructure of terror'. Human Rights groups such as Amnesty International were in no doubt, though, as to the real nature of the Israeli assaults:

> [T]he IDF acted as though the main aim was to punish all Palestinians. Actions were taken by the IDF which had no clear or obvious military necessity; many of these, such as unlawful killings, destruction of property and arbitrary detention, torture and ill-treatment, violated international human rights and humanitarian law. The IDF instituted a strict curfew and killed and wounded armed Palestinians. But they also killed and targeted medical personnel and journalists, and fired randomly at houses and people in the streets.[1]

These human rights violations aside, the IDF also directly targeted Palestinian Authority institutions, seizing their documents, hard disks and other records, and vandalising physical infrastructures, in what seems to have been, in some cases, a systematic attempt at de-institutionalisation.[2]

Numerous Palestinian NGOs, including human rights monitoring organisations, were ransacked, as too were private businesses.[3] Houses were demolished, health and medical services were incapacitated, roads were ripped up by the weight of columns of tanks, and in many places electricity and other vital services were paralysed. Curfews remained in place for upwards of three weeks; in Bethlehem the curfew lasted some 38 days. By the end of this period, the IDF had caused $361 million worth of physical damage – and this in an economy with an estimated GDP of only $4 billion.[4]

As with every other area of Palestinian life, water supply services were gravely affected. Pipes were ruptured by tanks and trenches, water spilling down the streets; pumping stations and wells ran out of diesel fuel or lost their electricity supplies; roof-top water tanks were deliberately shot at by Israeli troops; and under curfew, local engineers were often unable or too frightened to undertake necessary repairs. In Nablus, around 30,000 people went without piped water for 11 days in a row.[5] In Ramallah, at least 25,000 people lost their supplies for several days.[6] In Jenin, amidst the piles of corpses and bulldozed sewerage pipes, children screamed for water and drank sewage.[7] Oxfam estimated that, as of 4 April, 400,000 people in Ramallah, Nablus, Qalqilya, Bethlehem and Tulkarm were without access to running water.[8] The following brief snapshot from Nablus is indicative of what so many people experienced across the West Bank:

> Hajja Badriya had prepared for the expected invasion. She had stored extra water in every available container in the house. But she hadn't counted on the demolition of a nearby apartment building and its 15 homes on the fourth day of the invasion. When the residents fled the Israeli demolition, Hajja Badriya took in three more families. The additional people in her house quickly finished her carefully reserved water. They then resorted to searching the streets for pipes broken by passing Israeli tanks.[9]

In Tulkarm, a water engineer was shot and killed while trying to carry out repairs.[10] In Ramallah, the IDF entered the offices of the Jerusalem Water Undertaking (JWU), the water utility in the city, and arrested seven engineers and technicians who at the time were actively involved in co-ordinating repair work with the IDF.[11] Five of them were held for 18 days, before being released without charge.[12] The Local Aid Co-ordination Committee, which co-ordinates the international donor effort in the Palestinian territories, estimated that Operation Defensive Shield caused just under $7 million worth of damage to the West Bank's water and sewage infrastructures.[13]

The problems did not end there, however. In the months after Defensive Shield, the West Bank and Gaza were held in an ongoing state of siege. Checkpoints and barbed wire barricades were tightened still further. The Israeli Civil Administration in the West Bank introduced a new permit system, requiring Palestinians to get permission from the Civil Administration for travel between population centres.[14] Military incursions continued. During summer 2002, West Bank towns were held under almost continuous curfew. Inevitably, the economic and also the humanitarian situation in the Territories deteriorated to a new low.[15] Even before Defensive Shield, the Palestinian Authority had become 'effectively bankrupt'.[16] Israeli tax remittances to the PA – accounting under Oslo for about two-thirds of all PA revenues – had been withheld since December 2000, to the tune, during 2001 alone, of $0.5 billion.[17] Total tax revenues to the PA had dwindled by March 2002 to one-fifth of previous levels.[18] Half of the Palestinian population was estimated to be living below the poverty line, on under $2 per person per day.[19] Understandably, most Palestinians found water increasingly difficult to pay for, non-payment for piped supplies having risen in many areas to 80–90 per cent.[20] Municipal water suppliers were facing difficulties in turn, being forced to cut expenditures by withholding staff salaries, reducing services, and often leaving newly constructed and upgraded water networks unmanaged and unattended. In the 250 villages unconnected to the West Bank's water network, meanwhile (or connected to it, but nonetheless lacking regular or sufficient supplies) the situation was even more troubling. Not only were private tanker companies routinely being prevented, courtesy of Israeli closures, from supplying their water to these communities; just as seriously, most Palestinian families were finding themselves unable to afford the inflated prices for tanker-supplied water, of up to NIS 20 ($4) per cubic metre of water.[21] Palestinian communities across the West Bank were facing chronic water problems.

Not that the water situation in the Palestinian territories had been particularly rosy even before the breakdown of the Oslo process. Gaza's shallow groundwater resources had become heavily and increasingly saline – a result of over-extraction, and the consequent inflow of saltwater from the Mediterranean. In the West Bank, meanwhile, most Palestinian towns and villages suffered lengthy and severe water shortages throughout the Oslo period. Towns such as Bethlehem and Hebron faced regular water rationing, to the extent that, during summer 1998, most areas within Hebron received piped supplies for just one day in 20.[22] Parts of Bethlehem went without piped water during 1998 and

1999 for over three months.[23] And in most rural areas the situation was even worse, with villages such as Quasiba, mid-way between Bethlehem and Hebron, receiving no piped supplies for at least five months, from mid-April onwards. According to one influential study, West Bank and Gaza Palestinians were together receiving, in 1998, an average gross domestic supply of just 38 cubic metres per year (cmy).[24] If water losses are taken into account, net per capita municipal supplies in the West Bank averaged only 17 cmy.[25] To place this figure in some perspective, the internationally accepted 'minimum domestic water requirement' per person is commonly defined as 100 cmy.[26] Even before the breakdown of Oslo, the Palestinian territories had been in the midst of an ongoing water crisis.

THE PROMISES OF PEACE

Nine years previously, expectations had been so different. On 13 September 1993, Yasser Arafat and Yitzhak Rabin had famously shaken hands on the White House lawn. Israel and the Palestine Liberation Organisation had officially recognised one another, and had signed up to a short peace agreement, the Declaration of Principles, which established a framework and timetable for what would soon become known as the 'Oslo peace process'.[27] Much work doubtless lay ahead in resolving conflicting interpretations of, and in implementing, this Declaration, but it nonetheless seemed evident to the watching politicos and the mainstream Western press that peace between Israel and Palestine lay just around the corner. This was, as the beaming President Clinton pronounced with characteristic flourish, a truly historic moment, one that opened 'a new era not only for the Middle East, but for the entire world'.[28]

Within the month, international donors had pledged over $2 billion for the reconstruction and development of the West Bank and Gaza, in what was to become, in per capita terms, the largest donor effort ever undertaken by the international community.[29] Many areas cried out for attention, but one such was the water sector, which was viewed by many donors as a social, political and economic priority. It was widely recognised that the water situation in the Occupied Territories was increasingly critical, and that this problem needed to be speedily addressed, both so as to show the tangible benefits and rewards of peace (and thus build political support for the peace process) and also so as to improve the quality of Palestinian life, as well as to support economic growth and reconstruction. 'Peace, at last, now promises to provide the foundation for sustainable development in the Middle East,' proclaimed the then World Bank President Lewis Preston, and in line with this reconstruction and development-oriented thinking, the Bank placed water alongside transport as the sector most in need of investment.[30] The United States Agency for International Development likewise defined the water sector as one of its three 'strategic investment areas', and devoted to it the major part of its financial support to the Palestinians.[31] Led by these two donors, and with the support of many others, over 10 per cent of all aid money to the Palestinians between 1993 and 2000 went to water and wastewater projects.[32] Israel and the PLO had agreed in the Declaration of Principles to co-operate in managing the precious water resources of the West Bank and Gaza Strip. Given this, and with so much international support waiting in the wings, the clear promise and expectation was that the Palestinians of the West Bank and Gaza would soon be witnessing a marked improvement in the quality and quantity of their water supplies.

The questions of what happened to these unprecedented expectations, and of why, nine years later, Israeli troops were occupying Ramallah, Bethlehem and Jenin once again, and Palestinians were being forced to make do without adequate water supplies, lie at the very heart of this book. The answers I provide should be of interest both to followers of contemporary Israeli–Palestinian politics, and to students of Middle Eastern and international water problems. For those engaged and concerned with (or simply confused and perplexed by) recent developments in Israel and the Occupied Territories, this book provides an analysis of what went wrong. My approach, though, is a little different from the norm. Most analyses tend, for understandable reasons, to focus on the 'high politics' of the conflict – on troop withdrawals and redeployments, on negotiations between political leaders and their emissaries, on security co-ordination between Israeli, Palestinian and CIA officials, on settlement growth, suicide bombings, and so on. My approach, by contrast, is to focus in large part on a single issue, that of Israeli–Palestinian water politics, and to use this as a lens upon Israeli–Palestinian relations and the broader Oslo process. The tenor of my analysis holds much in common with those critical analyses developed by Noam Chomsky, Edward Said, Meron Benvenisti, Marwan Bashara, Graham Usher and others, all of whom, despite the differences in emphasis between them, view the Oslo 'peace process' as having been a disaster since its inception – a largely fraudulent means of maintaining and restructuring Israeli hegemony that was never likely to bring about a just or lasting resolution to the Israeli–Palestinian conflict.[33] Hopefully this book will complement and add further weight to the existing body of critical work on the Oslo process by providing a fine-grained account of why and how Israel has maintained its control over the shared Israeli–Palestinian water resources, and of why, nine years beyond the famous handshakes, the Palestinian populace of the West Bank and Gaza still found themselves in the midst of a severe water crisis.

For those directly concerned with Middle Eastern and international water politics, the analysis herein should be of interest in itself, as an empirically grounded case study in water conflict. Beyond this, though, I also seek to challenge some of the prevailing ways in which water crises and conflict in general are thought about – to provide a case study-based rethinking of a set of issues and problems which extend well beyond the Israeli–Palestinian arena. Most books on Middle Eastern and indeed international water politics have as their starting point the rivers, lakes and underground bodies of water (aquifers) that together constitute the hydrological foundation for social and political activity. This book, by contrast, starts off with individual people, communities and their water problems – their turned-off supply lines and their bullet-riddled water tanks – and seeks to enquire why these problems arose and why they are so difficult to resolve. My contention here is that water problems need to be understood, above all, with an eye to the broad political and economic contexts within which they emerge. Those who are familiar with existing work on Middle Eastern water issues may already have been wondering about the relevance of IDF human rights violations, or of the work of Chomsky and Said, to the analysis of problems that are discretely and specifically to do with water. To this I would simply say that it is not possible to explain water conflicts, either in the Middle East or elsewhere, except in relation to broader structures and relations of politics and political economy. For those engaged with water issues, this book seeks to illustrate why most of the prevailing technical and liberal work on the subject – work which tries to analyse water problems largely in isolation from their broader structural contexts – can only provide quite limited explanations of the frequently dire problems at hand.

THEORETICAL ORIENTATIONS

The approach adopted herein is based on what I hope are some well-founded general contentions about the nature of the social world, and about the possibility of it being adequately explained and understood. Thus before considering this literature in greater depth, I need to articulate something of my broad theoretical approach to social and political enquiry, specifically as it relates to the study of Israel, the Palestinians, and the international politics of water.

Firstly, I agree with Max Horkheimer, Immanuel Wallerstein and C. Wright Mills, amongst others, that socio-political analysis should be grounded in an understanding of the social world as a totality.[34] The modern social sciences are conventionally divided into discrete academic disciplines (economics, politics, law and so on), with each of these reflecting and focusing on a particular bounded domain or aspect of social reality (thus economists study market exchanges, political scientists study policy and government, and so on). Moreover, the disciplines themselves are often sub-divided into separate realms (say, between domestic politics and international relations), and thence into distinct 'levels of analysis' (for instance, between individual, state and inter-state systems, in the case of international relations).[35] Such is the standard positivist or empiricist mode of enquiry in the social sciences. By contrast with this, I assume that individual social phenomena can only be adequately explained and understood in relation to the structured social totality of which they form a part. The Israeli–Palestinian water conflict, for instance, can only be adequately explained when considered within the context of broader structures and relations – the nature of the Israeli–Palestinian conflict, the nature of Israeli and Palestinian economies, politics and societies, and their positions within an increasingly integrated 'modern world system'.[36] This water conflict can only be adequately explained, moreover, when viewed as a totality in which there are definite causal relations between its various economic, political, hydrological and other dimensions. Most works on water politics distinguish sharply between these dimensions, consigning 'economics', 'politics', 'hydrology' and so on to their own separate chapters. Such an approach tends, however, to inspire an elision of the issue of the causal relations between, say, economic and political facts, and tends to furnish accounts that are strong on description, but very weak on explanation. More will be said about this in the following section. Suffice to say now, though, that my aims in this book are above all explanatory, rather than descriptive; and that I thus try to analyse the relevant economic, political, hydrological and other details of this water conflict, not in isolation, but with an eye to questions of how they are causally connected, and of how they form a part of the structured totality that is the present-day modern world system.

Secondly, and following on from this, my explanatory claims are broadly historical materialist, concurring with Marx's famous dicta that 'social life is essentially practical,' and that 'it is not the consciousness of men that determines their being, but, on the contrary, their social being that determines their consciousness'.[37] Explanations of Middle Eastern politics and society often place emphasis, in Weberian fashion, on the determining causal effects of culture and value. It is often argued, for instance, that political structures and relations across the Middle East are deeply ingrained with the heritage of Islam – this being taken to explain the region's supposed traditionalism, its predominantly patrimonial patterns of rule, the weakly developed state of its civil societies, and its wholesale dearth of democratic governments, amongst other things.[38] I presume, by contrast, that such patterns of government and rule have arisen for practical and material reasons that have very little to do with Islamic ideas and principles, and that they do not differ in any fundamental way from those encountered elsewhere in the Third World. Thus stress is placed here, not on religion and culture, but on the legacies of colonial rule, on the uneven development of capitalism, on the distribution and control of economic and coercive resources, and on the organisational and practical difficulty of maintaining control (and constructing effective and legitimate state institutions) in the face of long-historical and structural constraints.[39] I seek to explain the Israeli–Palestinian water conflict, in short, not as the product of any cultural norms, values or ideas, but as a political economic problem which has its roots in patterns of capitalist development, and in concomitant patterns of state formation and state-society relations. While in subsequent chapters I make some use of non- and anti-Marxist theoretical resources – including formulations developed by Max Weber, Michel Foucault and Michel De Certeau, amongst others – I do so within a broadly historical materialist framework, and hopefully without too much conceptual confusion or theoretical inconsistency.

Third and finally, the implication of this Marxist focus on the practical and material roots of socio-political phenomena is that truth claims, too, must be viewed as forms of practice, ones that in large part reflect the practical purposes for which, and contexts in which, they are formulated and disseminated. Knowledge, to adapt the words of

Robert Cox, 'is always for someone and for some purpose'.[40] This does not mean that all knowledge is simply an ideological accessory to capital – when it comes to Israeli–Palestinian water politics, this is very far from being the case – and neither does it imply, in postmodern fashion, that truth claims can only be compared and evaluated on subjective, political or aesthetic grounds. What it does mean, however, is that all knowledge is interested, and is inevitably oriented towards certain (variable) purposes. Cox usefully distinguishes here between what he calls 'problem-solving' and 'critical' forms of knowledge.[41] The first of these is, as its name suggests, oriented towards problem-solving: it operates by slicing up the world into discrete spheres, and through so doing, by providing truths which are readily useful for policy-making, planning, administration and other instrumental purposes. The latter, by contrast, is oriented towards structural explanations, and out of that critique, assuming the social world to be a historically constituted structured totality that must be understood and explained as such. The pertinence of this in the present context is that, while most existing knowledge and literature about Middle Eastern water issues is of a broadly problem-solving character, my aims in this book are primarily critical. To clarify what I mean here, and to bring these comments somewhat back down to earth, we need to turn in more detail to the existing 'problem-solving' work on Middle Eastern water issues.

INSTRUMENTAL TRUTHS, LIBERAL INTERVENTIONS

An enormous amount has been written over recent years about the Middle East's water problems, conflicts and crises. Since 1993, at least seven volumes have been published providing broad surveys of the water politics of the Nile, Tigris-Euphrates and Jordan basins – these ranging from populist and sensationalist works such as Bulloch and Darwish's *Water Wars*, to comparative analyses such as Greg Shapland's *Rivers of Discord*, through to thematic works like Tony Allan's study of the relations between Middle Eastern hydro-politics and the global economy.[42] A further ten books have been written analysing various aspects of Jordan basin water politics, most of these with especial reference to the Israeli–Palestinian water conflict.[43] Then there are the edited collections, at least nine in recent years, with most focusing once again primarily on Israel and its Jordan basin neighbours.[44] Add to this the innumerable articles, papers, plans and policy proposals, and it becomes clear that there already exists a surfeit of information on the Middle East's water problems. What then remains to be said?

My blunt answer would be a great deal. To explain why this the case, though, we need to consider the question of the purposes for which knowledge about the Middle East's water problems are formulated. I have already suggested that most of such knowledge is problem-solving in character, being oriented towards specific reformist interventions. To put some flesh on this rather sweeping claim, it can be said that most of this information is produced by the state (or private sector consultants contracted by the state, and by inter-state organisations such as the World Bank), for state-centred governmental functions. Whether one looks at data on ground water levels, water quality, population growth, water consumption, or leakage from distribution networks, this data is largely produced by state agencies (or by other private sector or intergovernmental institutions linked to them), for state-led administrative and developmental purposes, and with the territory, population and economy of the state as its central unit of analysis. Qualitative claims about water issues are likewise disseminated primarily by the state, in the form of policy announcements, policy justifications and so on. At the centre of knowledge about Middle Eastern water issues, in short, lies the specific (though variable) institutional form that is the modern administrative, bureaucratic and governmentalised state.[45]

The significance of this epistemological predominance is that supposedly disinterested analyses of Middle Eastern water problems can all too easily become informed by state-led ways of defining and depicting them, often to the detriment of other issues and perspectives. This can happen in various ways, and for a variety of reasons. Experts and students have, in the first place, very little option but to rely to a large extent on state-oriented and state-formulated sources of information. When they want, say, data on domestic water consumption levels, they turn to government-produced planning documents, or to similar assessments by international and overseas development agencies. Equally, when they want the latest word on regional water conflicts, they quite reasonably seek interviews with relevant government officials. They are thus almost inevitably reliant upon the knowledge-producing organs of the administrative state. Beyond this, though, there is a complex sociology, politics and political economy involved in being 'an expert'. Leading water experts routinely serve as consultants to governments, to overseas development agencies and to international organisations like the World Bank – indeed their input into administrative, developmental and policy-making institutions is pivotal to their very status as 'experts'. Moreover, as is the case with experts in general, water experts typically operate with certain tacit ideas regarding what constitutes practical and policy-relevant knowledge, these tacit ideas being generally shared with governmental and governmentally linked water institutions. Most even 'independent' water expertise is thus in large measure shaped by and implicated in the specific apparatus of water-related 'power-knowledge', these existing on various scales (state, regional and global) but in each case revolving around state and state-linked administrative institutions.[46]

The consequences of this are several. Just as state-produced planning and policy documents address 'the water sector' as a discrete sphere of reality so, likewise, books and articles on the subject typically focus narrowly on water issues, without framing discussion of these in relation, for instance, to broader patterns of politics, political economy, state formation and development. Just as most administrative documents are highly quantitative and technically descriptive, aiming towards identifying specific technical solutions, so too most expert texts on water tend to privilege description and prescription at the expense of explanation and causal analysis. Furthermore, just as administrative plans standardly depict a world that consists only of impersonal objects of governance, largely devoid of living subjects – people – so also most books on water politics tend to be largely peopleless and disembodied (the exception here being well-known politicians, who often figure highly). Just as the formal faces of government documents always hide the political and personal struggles lying behind their eventual form, so likewise do most expert accounts skirt over conflicts within water institutions – observing the tacit rule that such matters should be discussed at international conferences, but not registered too tactlessly in print. Finally, water experts tend to see and present themselves as voices of liberal reason – what would be the point of their expertise if they did not? – but often in so doing, end up constructing a series of largely fictional irrational others which need to be overcome. Politics, society, culture and entrenched belief systems are regularly posited and portrayed as impediments to technical reason, and barriers to progress. In these various respects much of the existing literature on Middle Eastern water issues tends to mirror in its emphases and preoccupations, and also in its silences, those administratively useful problem-solving truths, produced by state and state-linked governmental institutions. Water experts and water institutions alike tend to see knowledge in instrumental terms, as useful to the extent that it serves to identify practical interventions. Equally, each of them typically sees the world through liberal lenses – as evidently amenable to the power of instrumental reason, modernisation and progressive development…if only all those irrational and traditional ways of thinking and operating, and all those power-driven politicians, could be marginalised and overcome.

This is, of course, something of a caricature. There are great differences in content, tenor and tone amongst the various works on Middle Eastern and Israeli–Palestinian water issues, with some being quite problem-solving in orientation, others much less so (these differences will be reflected on in more detail in subsequent chapters). Nonetheless, the existing literature on the subject does tend to repeat a series of standard themes, and tends to make broadly similar explanatory assumptions – in large part because of its close connections with knowledge produced by and for the problem-solving administrative state. As a result, this literature is marked by a whole series of blind spots. Little is said within it about structural contexts and constraints (since these offer no hope for technical interventions by water experts); little is said about internal water conflicts, or about the difficulties states have in controlling their domestic water arenas (states would much rather not publicise such issues); little, likewise, is mentioned of the uncertainties surrounding scientific and technical data (since these are too complex to discuss in any depth, and since states in any case would rather not undermine the epistemic foundations of their own planning and decision-making); and next to nothing is written about the micro-sociological dimensions of water crises (since, understandably enough, the business of administering water and of advising and intervening in water policy debates requires hard facts and figures, not problematically fluffy accounts of people's everyday social activity). Again, accounts do of course vary in their breadth and their specific emphases. Nevertheless I would contend that, despite the voluminous number of words that have been written in recent years on the subject of Middle Eastern water issues, a great deal remains unsaid. For followers of Middle Eastern and international water politics, this book attempts to say at least some of these things.

In constructing this study I have inevitably had to rely on many of the same sources of information as the existing problem-solving literature. However heterodox a political analysis may aspire to be, it still has to rely in large measure on knowledge produced for instrumental and administrative purposes – simply because problem-solving so dominates the epistemological grid of modern societies. Nonetheless, I have tried to escape from such sources to as great a degree as possible. The following chapters thus make use not only of conventional sources of information, but also of a range of sources that are sometimes hidden, and at other times just ignored: testimonies from local water engineers and administrators, to which international experts often get little access; unattributable gossip about delicate political issues; narrative accounts of people's everyday experiences of water crisis; technical documents in which political relations are discreetly hidden; and eyewitness evidence of people's coping strategies. No account can ever be wholly divorced from the power/knowledge circuits on which it has to rely. My hope is simply that by using a wide range of sources, I have managed to achieve a degree of critical distance, and managed in the process to construct an account that is both original in its perspectives, and not too tightly bound by the existing problem-solving orthodoxy. Foucault once defined critique as the practice of 'making facile gestures difficult', and while this is a little too dismissive, it is broadly in this spirit that the present book is written.[47]

STRUCTURE OF THE BOOK

The book is divided into two sections, with the first establishing some conceptual and historical parameters for the subsequent (and more thoroughly empirical) analysis of Israeli–Palestinian water politics during the Oslo period.

The first two chapters focus in turn on the causes of water crises, and the causes of international water conflicts. Chapter 1 develops a typology of some of the different ways in which water crises have been characterised, explained and understood by water experts and institutions, and via a critique of two of these types of explanations – specifically, those inspired by Malthusian and by liberal reasoning – ends up defining water crises in general as problems of 'techno-politics' and political economy. This chapter develops and substantiates many of the critical remarks already made about problem-solving knowledge. Chapter 2 turns to questions of international relations, considering contending understandings of the Jordan basin's twentieth-century history of water conflict. Critiques are developed of liberal and realist explanations of Jordan basin water politics, the one for its political naïvety, the other for its elision of political economy and its insensitivity to history. Transcending these limitations, I offer as an alternative a Marxist-informed account of the historical roots of Israeli water policy, one that provides a long-historical explanation for Israel's ongoing water conflict with its Jordan basin riparians and with the Palestinians. Chapter 3 further develops this narrative by focusing on Israeli–Palestinian water relations during the period of direct occupation, 1967–93. Read together with the latter part of chapter 2, this provides the necessary historical backdrop for understanding Israeli–Palestinian water politics under Oslo.

The second part of the book analyses these post-1993 water relations. Chapter 4 is largely descriptive, serving to identify the effects that the Oslo process had on Israeli–Palestinian water relations – my main contentions being that these changes were in many respects more discursive than real, and brought few new benefits to the Palestinians. Chapter 5 digresses briefly into the contested politics of hydrology in the Israeli–Palestinian arena, examining the strong possibility that the one additional groundwater resource made available to the Palestinians as the result of the Oslo agreements might already be being exploited at unsustainable levels. In chapter 6, we return to more straightforwardly political matters, by considering the explanatory questions of why the changes (or lack thereof) described in chapter 4 came about, and of why the Oslo process broke down. Answering these questions involves focusing above all on the Oslo negotiations, and the various motivations and power relations that informed them. Chapter 7 turns to the 'internal' Palestinian arena, considering some of the problems encountered by the Palestinian Authority in administering and developing its water sector under Oslo, including its relations with international donors and contractors. Finally, chapter 8 is ethnographic, providing a sympathetic (and, dare I say it, almost hopeful) account of the everyday coping practices of Palestinian water users in the southern West Bank. Each of these chapters attempts not only to tell its own empirical tales about the Israeli–Palestinian water conflict, but also to use these as a lens upon the Oslo process in general, and upon expert discourse on Middle Eastern and international water problems. A brief conclusion returns us to Sharon and his tanks, and also attempts to draw out some of the book's several practical and policy implications.

A final introductory note: most of the analysis herein focuses primarily on the West Bank and, more especially, on the southern West Bank areas around the Palestinian towns of Bethlehem and Hebron. I say very little about the quite different and, in ecological terms, much more precarious water situation in the Gaza Strip. Equally, I say very little about water quality issues, focusing above all on issues of water supply – the former being much more acute in Gaza, the latter more pressing in the West Bank. It should be emphasised that these omissions are not founded in any sense that Gaza or the northern West Bank are unimportant as far as water issues are concerned, but merely reflect the primarily explanatory and interpretive, rather than descriptive, aims of this study. Throughout I have tried to analyse 'water politics' at many different levels, and to be as thematically comprehensive as possible. Inevitably something has had to give way – and that something, in the present case, is descriptive and especially geographical breadth.

PART ONE

Theoretical and Historical Contexts

CHAPTER 1

Explaining Water Crisis: Ecological, Technical and Political Discourses

Critical water problems exist across the world – at a range of scales, at a range of intensities, and involving a range of specific issues, actors, contexts and circumstances. In Iraq, water and sanitation facilities have remained in a continuing state of disrepair ever since the Gulf War of 1990–91, with clear impacts on the incidence of water-borne diseases, infant mortality and malnutrition.[1] In south-eastern Turkey, Kurdish villages continue to face the threat of destruction at the hands of the Turkish government's Ilisu dam project, while in the north-east of the country, 15,000 ethnic Georgians face the threat of displacement from the planned Yusufeli dam.[2] In Dhaka, capital of Bangladesh, the Buri Ganga River serves as a multi-functional transport route, religious site, wash house, rubbish dump and open sewer; the river regularly floods, and thousands die each year from water-borne diseases as a result.[3] In southern Africa, an estimated 13 million people (primarily in Zimbabwe, Malawi, and Zambia) faced crop failures and extreme food shortages during 2002, these having been brought on, at least in part, by recurrent droughts.[4] In Ghana, Bolivia, South Africa and many other countries besides, local conflicts are simmering over proposed and on-going privatisations of public water utilities to Western multinationals.[5] The list could go on and on.

The question that concerns me here, though, is how these disparate problems can best be explained. These problems each have their own particular causes, of course, and their own unique casts of characters, such that a fine-grained empirical account could readily be written about each of them. But this would not necessarily help us in understanding the general causes of water problems. Empirical work always carries with it the risk of sacrificing analysis for detail, and of failing to see the wood for the proverbial forest of trees. Empirical analysis can often leave the big questions unanswered, even ignored. Furthermore, even if we had a full set of highly persuasive accounts of these various water problems and crises, the question would still remain of what, if anything, links them together. What, we may ask, are the common factors behind these individual crises? What can be identified as the general causes of water crisis, both across the Middle East and beyond?

In trying to answer these taxingly broad questions, this chapter develops an analytical framework for explaining water crisis. My aim here is not to provide firm answers, but to clarify the terms of the question, and to articulate some of the conflicting ways in which water crises can be understood. There are, I suggest, three different ways to explain water crises, and three different discourses of water crisis – one ecological, one technical, and the other political in emphasis. Each of these three discourses presents, and represents, a particular take on the nature and causes of water problems, and each of them also enshrines its own particular perspective on the most appropriate responses to, and the most likely outcomes of, water crisis. Each, moreover, is premised on a set of assumptions about the nature of the human condition, and about the character of human relations with the natural world. Each of these three discourses implies and carries the marks of a distinct logic, a distinct worldview.

These three discourses are, of course, artificial analytical constructs, ones that in their 'conceptual purity... cannot be found anywhere in reality'.[6] No single author or institution espouses purely ecological views, or argues from a wholly technical or political perspective; and while water experts frequently differ in their assessments of the causes of water problems, these differences are rarely absolute, and are never confined within ideal-typical trenches. These discourses are not 'regimes of truth' that imprison their authors and readers within limited conceptual horizons.[7] Quite the contrary, differences between water experts are always a matter of degree and overall emphasis. My purpose in distinguishing between these discourses is simply to clarify some of the assumptions on which various actual accounts are founded – and in so doing to pave the way for the more in-depth analyses in the chapters to follow. For those unacquainted with Middle Eastern or indeed international water issues, this three-way typology also serves as something of an introduction to existing perspectives and debates.

I begin by characterising each of the discourses, and identifying some of the contrasting assumptions on which each of them is based (these are summarised in Figure 1.1). In each case, I make both general comments about how the discourse in question explains water crisis, and more specific comments about its explanation of the Israeli–Palestinian case. Thereafter I develop critiques first of ecological and then of liberal technical discourse on the causes of water crisis.

Discourse	Problems	Solutions	Likely outcomes
Ecological	Scarce or finite resources plus high populations	Limit population growth	Water wars
Technical	Mismanagement and inefficiencies	Improve management and efficiency	Progress
Political	Uneven distribution of power and resources	Reduce power and resource inequalities	Winners and losers

Figure 1.1 Three Discourses of Water Crisis

THREE DISCOURSES OF WATER CRISIS

Ecological Discourse

Approached from an ecological perspective, water crises arise above all from the fact that, while already-high populations are inexorably rising, natural supplies are limited and constant. Malin Falkenmark and Jan Lundqvist's words are typical in this regard:

> [T]he available water resources are simply limited. In principle, the amount of water that can be made accessible for various purposes and to various groups of people is determined by precipitation falling over the catchment areas where water is required. The amount varies from year to year and between seasons. Apart from these fluctuations, which are quite significant in the South and may extend over long periods, the amount is virtually fixed. It is at the same level as it has been historically, and future generations will have to make do with the same amount. On a per capita basis, the availability is thus reduced in proportion to population increase.[8]

The guiding assumption here – characteristic of ecological discourse as a whole – is that water crisis is fundamentally a product of overpopulation relative to the available resources: as populations grow, so the finite resource base becomes more and more stretched, and so crisis ensues. 'Unfortunately, water resources are finite; future increases in population therefore imply increased water competition.'[9] This happens within particular regions, locales and states, but also, so some claim, on a global scale. Thus globally, world population growth is 'outrunning water supply', while the Middle East as a whole 'is "close to the ceiling" in terms of its very high number of people per flow unit of water'.[10] Across the Middle East, individual states are hitting a 'water barrier'.[11] And this is centrally because of imbalances in the population-natural resource equation.

From an ecological perspective, the straightforwardness of this population-resource equation is such that the depth and severity of water crises can be quantified and compared numerically. Malin Falkenmark, for one, has developed a 'water stress index', which aspires to do precisely this. In Falkenmark's terms, states with a per capita natural water resource availability of less than 1600 cmy can be thought of as exhibiting 'water stress', while those with an availability of less than 1000 cmy per capita face 'chronic water scarcity'.[12] By this rubric, Jordan, Israel and the West Bank and Gaza are all already in a state of 'chronic water scarcity'. To explain why this is so, one needs only to consider some brute ecological facts about the natural resource base of the Jordan basin states, and about the population loads that they have to bear.

The Jordan River (Figure 1.2) rises in southern Lebanon, northern Israel and the Golan Heights, from where it flows southwards to Lake Tiberias (otherwise known as the Sea of Galilee or Lake Kinneret), and beyond to the Dead Sea. The Jordan is joined by two main tributaries: the Yarmouk, which rises in Syria, and flows along the Syrian–Jordanian border before joining the River Jordan just south of Lake Tiberias; and the Zarqa, rising in Jordan. In addition to these, the Jordan receives water from numerous small and seasonal tributaries in Israel, Jordan and the West Bank, as well as from countless surface and underground springs. At its terminus lies the Dead Sea, over 400 m below sea level, which has no natural outflow and loses its water solely through evaporation. This river system, famous as it is, is absolutely tiny, discharging a mere 1.6 bcmy: compare this with 83.6 bcmy discharged by the Nile, and 81 bcmy discharged by the Tigris and Euphrates.[13]

These surface waters aside, Israel, Jordan, and the West Bank and Gaza also attain much of their water from underground 'aquifers', this

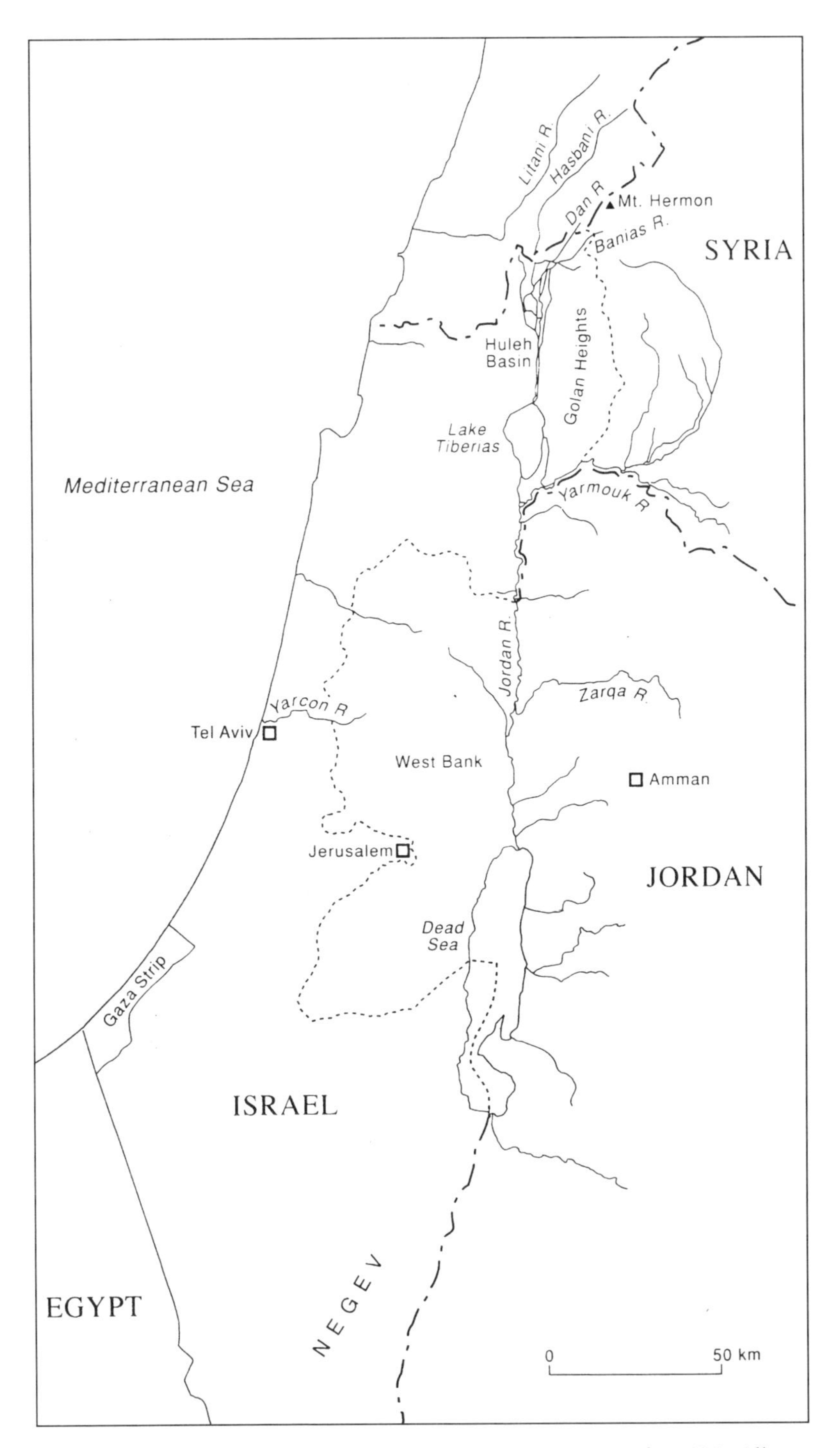

Figure 1.2 The Jordan Basin (reproduced, with permission, from J.A. Allan [ed.], *Water, Peace and the Middle East* [1996], courtesy I.B.Tauris Publishers)

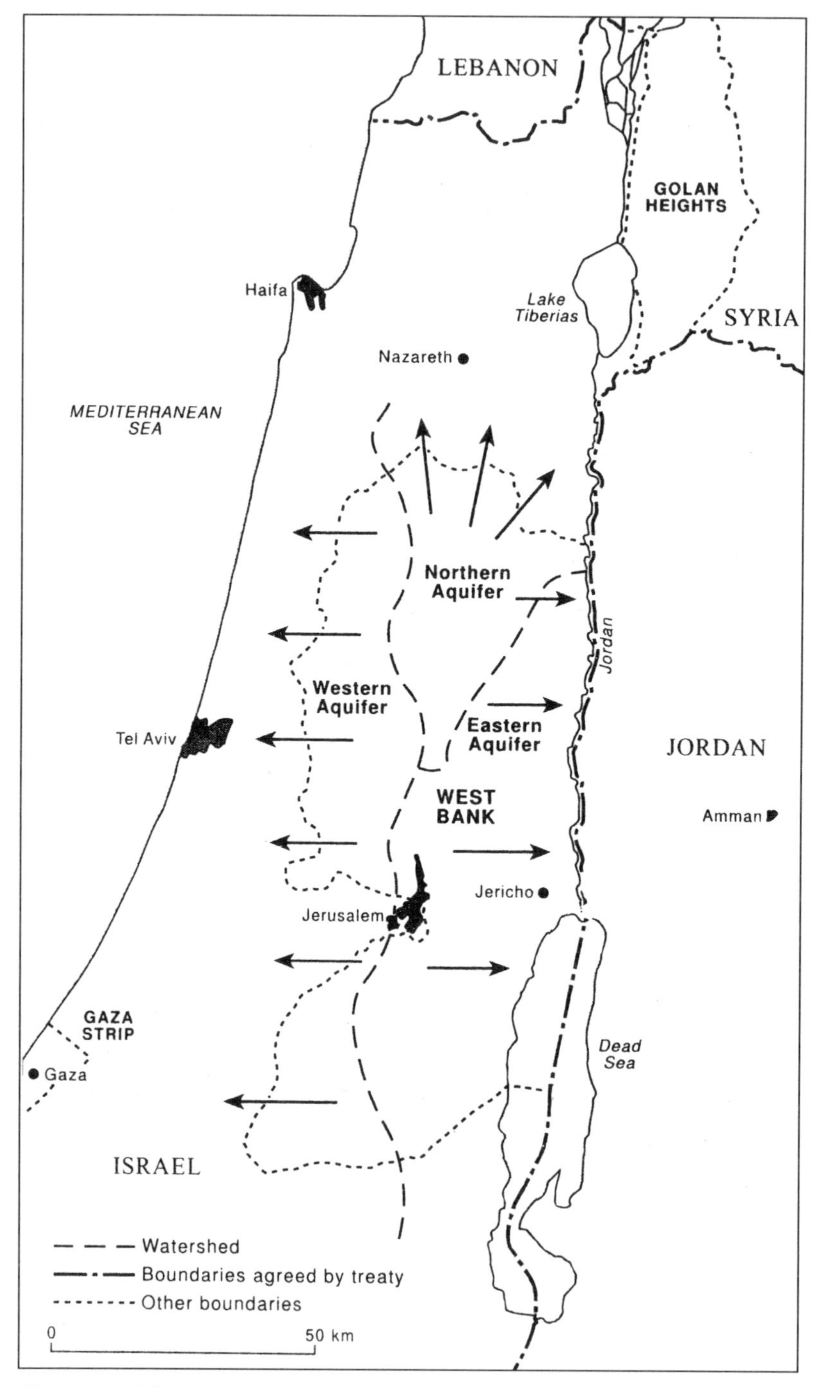

Figure 1.3 The West Bank Aquifers (reproduced, with permission, from Greg Shapland, *Rivers of Discord: International Water Disputes in the Middle East* [1997], courtesy C. Hurst & Co. [Publishers])

water being either discharged from springs, or extracted through often deeply drilled wells. Much of northern Israel and the West Bank, in particular, are home to heavy winter rainfall, and while some of this water flows over the surface, forming countless seasonal streams, much of it filters into the bedrock. From there, most of the groundwater flows westwards towards the Mediterranean (though a smaller amount flows east towards the Jordan River). Most of the rain falling on the West Bank flows underground into Israel, where it emerges naturally from springs at the edge of the country's coastal plain (Figure 1.3). Likewise – though here we are talking about much smaller quantities – the Gaza Strip's groundwater resources are in part fed with water filtering underground from southern Israel. The region's aquifers provide Israel, Jordan, the West Bank and Gaza with a renewable resource base of around 1.4 bcmy.[14] Overall, a mere 2.7 bcmy of water is naturally available to Israel, Jordan, the West Bank and Gaza from renewable surface and groundwater sources.[15]

Israel, Jordan, the West Bank and Gaza, in sum, have only very meagre natural water resources. What's more, their populations are high and growing. Israel numbers over six million people, and during the last decade has had an annual population growth rate of over 3 per cent (this being largely because of high levels of immigration, and despite a relatively low natural growth rate).[16] The West Bank and Gaza have an estimated Palestinian population of around three million, and are witness to a high natural growth rate of 3–4 per cent.[17] Jordan's population is around five million, again with a growth rate of over 3 per cent a year.[18] Taken together, Israel, Jordan, the West Bank and Gaza thus have a renewable water availability of less than 200 cmy per person – and hence are clearly, in Falkenmark's terms, in a situation of 'chronic water scarcity'. Viewed ecologically, Israel, Jordan, and the West Bank and Gaza are already exceeding the 'carrying capacity' of their collective, and extremely shallow, resource base. From such an ecological perspective, supply shortfalls and the over-exploitation of the region's surface and sub-surface resources – seen most distressingly in Gaza, where over-exploitation has rendered the shallow coastal aquifer increasingly saline – are the inevitable consequence of resource limitations combined with population pressure.

Add to this population-resource equation the assertion, often made by water experts, that economic development inevitably leads to increased per capita water demand; and add to this the claim, likewise often made, that global climate change is likely to increase the regularity of droughts, and one has nothing less than a recipe for disaster:

the future, from an ecological perspective, cannot but appear grim.[19] Human populations and their ever-rising demands are bound – sooner or later, and in more and more regions of the world – to run up against the absolute limits of nature. Once this happens, water is likely to become a constraint on economic development, and also a focus of political tension and inter-state conflict.[20] Water will begin to play not only a determinate role in shaping the economic and political futures of the Middle East, but also in causing violent conflict.[21] This is the case in the Palestinian territories as elsewhere; indeed as one leading analyst put it in a submission to the US Congress:

> If the [water] crisis is not eased, it will result in a significant rise in the probability of warfare... It is water, in the final analysis that will determine the future of the occupied territories... and by extension the issue of conflict or peace.[22]

In the face of these threats, the only meaningful policy option from an ecological perspective is to reduce population growth. Water resources, after all, are scarce, fixed and finite, and hence it is human over-consumption, and above all population growth, that most demand attention. 'A crucial step in averting a water crisis is to slow the unprecedented population growth occurring in so many of the world's countries,' argue Falkenmark and Widstrand, carrying the logic of their argument to its necessary conclusion. Similar is the position recently enunciated by the influential Washington-based Worldwatch Institute:

> If the world could move from the UN medium population projection of nearly 9 billion in 2050 to the low projection of less than 7 billion, water stress would be greatly alleviated, making the water problem much more manageable. If the world stays on the current population trajectory, a growing share of humanity may simply lack the water needed for a decent life.[23]

Ecological discourse – as should now be clear – offers an extremely pessimistic reading of water crisis.

Technical Discourse

Technical discourse, by contrast, is remarkably upbeat, foreseeing numerous ways in which water shortages and water quality problems might be addressed, ameliorated and overcome. From a technical perspective, water crises are above all ones of technological, economic and policy mismanagement and inefficiency. The words of World Bank

Vice-President Ismail Serageldin neatly encapsulate this position: 'the water problems in most countries stem mainly,' he contends, 'from inefficient and unsustainable use of water'.[24]

There are a number of variations on this general theme. For some water experts, technological underdevelopment and under-investment are the central reasons for concern, and hence water problems should be addressed through the development of appropriate material technologies – dams, pipelines, new distribution lines, desalination plants, wastewater treatment plants, drip irrigation systems, and so on. For others, water crises arise essentially out of the failure to treat water as an economic commodity, out of a failure to understand that 'water is mainly money'.[25] From such a perspective, water scarcity is not problematic in and of itself; scarcity, to the contrary, is a necessary condition of economic activity, without which there would be no need whatsoever for economic exchange. The problem, claim those of an economic bent, is that water is so routinely undervalued and often, as a result, allocated in an economically inefficient manner; hence the key to resolving water problems lies in the development of appropriate pricing and tariff systems, and also (at least for neo-liberal economists) in the privatisation of water resources, and the development of domestic and international water markets.[26] Still other water experts emphasise, not so much technological and economic inefficiencies, as administrative, institutional and policy-making ones – concurring with former World Bank head Robert MacNamara's claim that 'management is the gate through which social, political, economic, technical change ... is rationally spread through society'.[27] Some have averred that certain societies and polities are characterised by a lack of 'adaptive capacity', by a relative underdevelopment of appropriate social resources for coping with water scarcities.[28] Some, moving into yet more complex territory, hold that political issues are often ignored within technical analyses, and that assessments of the viability of technical proposals should include analyses of factors such as the power and interests of various parties.[29] There are great differences in emphasis here between these various technical assessments. Nonetheless, all such technical discourse places a common stress on mismanagement and inefficiency, and maintains a progressivist faith in the possibility of improving – and indeed optimising – water resource management.

The World Bank's writings epitomise this technical approach to water crises. For the Bank, water crises can be overcome so long as governments (in 'partnership' with consumers and also external sources of financial and technical support) make the necessary institutional, technical economic and policy adjustments. The following, for instance, are listed as priorities for the Middle East and North Africa region,

> *mobilizing governments and peoples* to promote national and local partnerships and participatory approaches for using water wisely; *integrating water resources management* to reconcile competing claims on limited supplies; *using water more efficiently* to get the most value from it; *seeking alternative sources of water* to free countries from reliance on finite freshwater; and *promoting regional and international partnerships* to foster technical and financial cooperation on water issues.[30]

If such measures were undertaken, maintain the Bank, the 'vicious circle' of regional water crisis, in which 'harsh water shortages adversely affect economic growth' could be transformed into a 'virtuous circle' of 'growth for water', whereupon development would enable the region to climb out of water crisis.[31] This applies to the Jordan basin states as much as in any other. Thus in the Occupied Territories the inefficiencies are first of all institutional:

> The institutional capacity of the municipal water departments is generally very weak. This is due, firstly, to the organizational structure of the sector and, secondly, to human resource constraints...The main institutional issues facing the OT infrastructure sector are the overly complex legal base, the use of municipal departments as the organizations responsible for service delivery, the lack of agencies dealing with the planning and coordination of activities in water supply and sewerage, inadequate managerial capacity, bureaucratic constraints and the difficulty in raising funds for investments and rehabilitation works...Shortages of professional and technical skills are a constraint to the performance of the sector.[32]

But the root problems are also technological:

> Existing supply networks are generally old; unaccounted-for-water frequently exceeds 50 per cent; meters are commonly inaccurate, broken, or bypassed; supplies are inadequately chlorinated; intermittent supplies and low pipe pressure cause reverse flow into the network and contamination; and water departments are generally weak and under-funded.[33]

In both institutional and technological terms, the Palestinian water sector is marked by a set of inefficiencies, by a series of 'constraints', 'lacks' and 'inadequacies'. For the World Bank, the challenge is one of turning these absences into presences, and of transforming an undeveloped water sector into a developed, well-managed and consequently efficient one. In stark contrast with ecological discourse, which posits development (and its attendant population growth) as the cause of environmental crisis, technical discourse presents development and modernisation as the only way out of crisis. Once these are undertaken, water crisis will be overcome. Such is the technical perspective on water crisis.

Political Discourse

From a political perspective, finally, water crises are essentially the product of inequalities, differences and conflicts. The words of Amartya Sen provide a useful entry point here: as he observed in his famous treatise on poverty and famine, 'scarcity is the characteristic of people not having enough...it is not the characteristic of there not being enough. While the latter can be the cause of the former, it is one of many causes.'[34] The question of whether resources (natural or technical) are scarce or not is, from this perspective, often quite beside the point; what matters much more crucially is how such resources are allocated and distributed in relation to human needs. From a political perspective, it is precisely the uneven distribution of resources that lies at the heart of the world's various water crises.

There are many different slants that one might take here. One might want to emphasise, in the manner of dependency and world systems theory, the structured relations between North and South that bind most of Africa, south Asia and Latin America into an ongoing state of dependency upon the capitalist core of the world economy – and the way in which water crises inexorably follow from this globally structured inequality.[35] One might want to stress, in the manner of much nationalist and national security discourse, the contrasting resource bases of individual states and nations – interpreting water problems as resultant from the uneven distribution of natural resources and national capabilities between neighbouring states. Alternatively, one might want to focus on those inequalities, tensions and conflicts that are internal to states – on the differences that lie between urban and rural areas, between individual communities, or between classes, genders, or even households and families. Political discourse can take various forms. What essentially characterises it, though, is an emphasis on differences, inequalities and conflicts between social groups – in whatever particular form these groups happen to take.

Political discourse on the Jordan basin centres almost entirely on the unequal distribution and control of water resources between neighbouring nations and states. The Palestinian water crisis from this perspective follows neither from any breaching of local resource thresholds, nor from mismanagement and inefficiency, but rather from Israel's control of regional water resources, from its denial of Palestinian water rights, and from the differential allocation of water between Israeli and Palestinian users. Jad Isaac is one of the most vociferous exponents of this perspective: 'in reality,' he asserts, 'the water crisis is not chiefly one of insufficient supply, but of uneven and inequitable distribution'.[36] Seen in this light, the key point here is that the Palestinians have only minimal control over and access to regional and local water resources. For example, while the West Bank borders the Jordan River, the Palestinians living there are effectively denied access to it, since Israel (and to lesser extents Jordan and Syria) use most of its waters before they even reach the West Bank: by the time it flows past the Palestinian town of Jericho, the Jordan River is so saline as to be unusable. The majority of the West Bank's groundwater resources are likewise used by Israelis rather than by Palestinians: at the time of the 1995 Oslo II Agreement, 85 per cent of the West Bank's groundwater resources were consumed by Israelis, and only 15 per cent by Palestinians.[37] The effect of this was (and continues to be) that per capita gross domestic supplies in Israel were three times what they were for West Bank Palestinians (100 cmy compared to 38 cmy in the West Bank in 1995) – and this disparity becomes even starker when one takes into account that water networks in the West Bank have much higher losses than those in Israel.[38] It is on the strength of facts such as these that political accounts of the Palestinian water crisis are founded.

Viewed from this perspective, the Palestinian water crisis will only be resolved through Israeli recognition of legitimate Palestinian water rights, and through Palestinian receipt of (and also control and ownership of) their rightful share of regional water resources. There are, predictably, a wealth of conflicting ideas as to what constitute 'Palestinian water rights', and as to how the region's waters should most equitably be redistributed (no doubt this variability in part reflects political partisanship, but it also arises from the fact that the central principles of international water law often conflict with one another, thus rendering it impossible to determine absolutely rightful water allocations).[39] For some, the primary frames of reference are the 1907 Hague Regulations, the 1949 Geneva Convention and the UN General Assembly's various resolutions on Israel and the Occupied Territories which, taken together, clearly indicate that Israel's exploitation of the natural resources of the West Bank and Gaza contravenes international law.[40] Others advance specific calculations of Palestinian water rights. Zarour and Isaac, for instance, argue for a water-sharing formula based on 'nature's apportionment', the impact of which would be to grant the Palestinians 80 per cent or more of the West Bank's waters, as against the 15 per cent that they received in 1995.[41] Others develop their proposals not on the basis of the natural distribution of water resources, but in relation to social and economic needs, with Shuval, for instance, favouring a resolution that respects annual per capita 'minimum water requirements', calculated as 100 mcy for all municipal (i.e. domestic, urban and industrial) uses.[42] Others argue, perhaps more in keeping with the principles of international water law, that both natural and social factors need to be considered in any calculation of Palestinian water rights.[43] Yet irrespective of the important differences between these various proposals, all of them assume that the Palestinian water crisis will only be resolved through a redistribution of regional and above all Israeli–Palestinian water resources. Such is the political perspective on the Palestinian water crisis.

To summarise: from an ecological perspective, water crises in general and the Palestinian water crisis in particular should be understood as functions of an imbalance between population and natural resource levels, of the fact that ever-expanding populations are dependent upon scarce and finite stocks of water. From such a perspective, water crises should be addressed through limiting population growth – and given the slim chances of achieving this, the future looks bleak, with de-development and water wars looming large on the horizon. From a technical perspective, by contrast, the key problems are ones of mismanagement and inefficiency. Given this, governments should respond by updating and indeed optimising technological, institutional, regulatory and pricing systems – and so long as they do this, water crisis can be resolved. Finally from a political perspective, water crisis follows above all from unequal access to, and control of, water and other related resources. Crisis could be addressed through the granting of water rights, and the re-allocation of water resources. Until this happens, however, the inequalities will inevitably continue.

Each of these three discourses embodies particular conceptions of the nature of water crises, of the necessary responses to such crises, and of their most likely outcomes. Each of them provides (to some degree) a plausible account of the Palestinian water crisis. The question that needs addressing, however, is whether they are all equally valid as explanations not just of the Palestinian water crisis, but of the phenomenon of water crisis in general. Which, if any, provides the best overall explanation for water crisis? Alternatively, can these three discourses be meshed together without contradiction within a single account? It is to these questions that we now turn.

THE PRODUCTION OF NATURE[44]

Ecological accounts of water crisis locate their causes, as we have already seen, in the seemingly inexorable clash between limited water resources and ever-increasing demand. This perspective, most commonly espoused by ecologists and hydrologists, arises from, and also needs to be understood within the context of, a broader Malthusian tradition of thinking about the relations between populations and resources. Malthus famously asserted in his 1798 *Essay on the Principle of Population* that while populations necessarily follow a geometric (i.e. exponential) rate of increase, the means of subsistence can only ever increase arithmetically; thus populations will sooner or later outstrip their available food supplies, and a state of war, poverty, famine and disease will ensue.[45] The fine detail of this argument was no doubt entirely spurious, but the concerns voiced by Malthus have continued to attract a following, especially during the past several decades, and given the ever-mounting evidence of human damage to, rootedness within, and dependence upon the biosphere. Hence the assertion, made most famously by the Club of Rome, that there are natural 'limits to growth'; hence the claims, often made, that we collectively face (or are already in the midst of) 'population explosions'; and hence the view, commonly held by many radical ecologists and environmentalists, that 'sustainability and industrialism are mutually exclusive'.[46] It is in relation to this broadly Malthusian tradition that ecological discourse on the Middle East's water crises should be understood. Thomas Stauffer puts the point especially bluntly, 'The Malthusian specter,' he observes, 'is real in the Middle East'.[47]

From this Malthusian perspective, natural resources impose absolute limits on human societies and economies, and we transgress these limits only at our peril. It is on the back of such assumptions that ecological discourse makes such frequent reference to 'thresholds', 'red lines', 'carrying capacities' and 'water barriers', each of these idiomatic expressions being suggestive of unalterable natural limitations on human exploitation and activity. Nature and society, from this perspective, are ontologically distinct and separate. Humans use nature – human encounters with nature being above all ones of 'use' – and nature imposes constraints upon this use, but beyond that the two are barely connected. Such, in ideal typical terms, is the Malthusian or ecological portrayal of nature and society.

Malin Falkenmark's claims are, as we have already seen, quite typical in this regard: for her, 'the available water resources are simply limited,' and on 'a per capita basis, the availability is hence reduced in proportion to population increase'.[48] Yet Malthusian assumptions also inform, albeit in a limited and no-doubt barely intended way, most expert analyses of Middle Eastern water problems. Take, for instance, Aaron Wolf's *Hydropolitics Along the Jordan River* – a book which is in many regards excellent, and which merely serves here as a representative example. Wolf begins his analysis by detailing the Jordan basin's hydrography:

> Natural System: Surface Water
>
> The Jordan River watershed drains an area of 18,300 square km in five political entities – Lebanon, Syria, Israel, Jordan, and the West Bank.
>
> Three springs make up the northern headwaters of the Jordan: the Hasbani, rising in Lebanon with an average annual flow across the border of 125 mcmy, the Banias in the Golan Heights, averaging 125 bcmy, and the Dan, the largest spring at 250 mcmy and originating in Israel. The streams from these springs converge 6 km into Israel and flow south to the Sea of Galilee at 210 m below sea level.
>
> The Yarmouk River has sources both in Syria and Jordan and forms the border between those countries before it adds about 400 mcmy to the Jordan, 10 km south of the Sea of Galilee. Beyond this confluence, the Jordan picks up volume from springs and intermittent tributaries along the 320 km meander southward along the valley floor of the Syrio-African Rift. At its terminus at the Dead Sea 400 m below sea level, the Jordan River has a natural annual flow of 1,470 mcmy ...[49]

Wolf then provides equally brief accounts of the Jordan basin's 'natural' groundwater systems, and of water use levels within the Jordan riparian states; and then offers an exposition of the history of conflict and cooperation over the Jordan basin. In subsequent chapters, Wolf addresses the technical, economic, institutional, political and legal dimensions of water crisis, and investigates the potential utility of various 'paradigms' in promoting solutions to water crises, such that Wolf's account makes use of a wide variety of disciplinary perspectives. Yet there is a covert Malthusianism lurking here. Nature (the resources) and society (human use, history, and so on) are presented as entirely separate, and wholly antithetical. Nature is portrayed as the firm foundation and timeless stage upon which the superstructure of society is erected and enacted. 'Natural systems' are implied as being completely different from those 'new sources' of water capable of being produced 'through technology' (whereas the one constitutes an analytical starting point, the other lies only in the future).[50] Natural water resources are passive and calculable, unchanging and devoid of history – characterised, in short, by everything that human society is not.

Most accounts of Middle East water issues are broadly similar to this in their assumptions: they virtually always begin by itemising and profiling the natural water resources, only afterwards surveying the technical, economic, political, legal and so on dimensions of water crisis. This narrative mode is in keeping with the generally problem-solving character of work on Middle Eastern water issues. Yet for explanatory purposes, the claimed existence of natural water resources is problematic in at least two regards. To start with, the positing of a pristine and pre-human nature ignores the incontrovertible fact that our present-day 'nature' is thoroughly 'humanised'.[51] Marx and Engels recognised this even a century and a half ago, observing that 'the nature that preceded human history ... is nature which no longer exists anywhere (except perhaps on a few Australian coral islands of recent origin)'.[52] If we add to this a further 150 years of economic development and environmental transformation, and also the mounting evidence of human-induced atmospheric and climatic changes reaching into every corner of the earth, it can quite reasonably be asserted that we have now witnessed the 'end of nature'.[53] The Jordan basin is a case in point. For while Wolf writes that at 'its terminus ... the Jordan River has a natural annual flow of 1,470 mcmy,' the socialised reality is to the contrary that only 152–203 mcm of water flow into the Dead Sea each year – most of the Jordan basin's waters being diverted by Israel, and to lesser extents Syria and Jordan, towards the capital cities of Tel Aviv, Damascus and Amman.[54] Likewise, the Jordan basin's aquifers have all witnessed declining water table levels and increased salinity over the past half century, and, in various complex and often unclear ways, have all been transformed by changing patterns of urbanisation, deforestation and agriculture (since these each have effects upon patterns and rates of infiltration into the aquifers). Across Israel and the Palestinian territories, land use changes have brought with them changes in rainfall patterns, with rainfall in the Negev Desert, for instance, having increased by up to 30 per cent in recent years.[55] It has long been part of modern Western imagery, of course, that nature is ontologically separate from, and prior to, society.[56] However, like all else in our biosphere, the Middle East's 'natural' water resources are in truth completely saturated with the marks of human labour.

A second problem with the quasi-Malthusian assumptions of most accounts of the Middle East's water crises is evidenced, for instance, in Wolf's distinction between 'natural systems' and 'new sources through technology'. Descriptively innocuous as this distinction undoubtedly is, it does nonetheless elide the fact that all water resources are produced and rendered useful through technologies, even when they are thought of as being natural. Without the use of dams, pipelines, pumping and storage systems, it would be impossible to collect, control and make available the waters of the Jordan River. Without the existence of complex drilling and pumping technologies capable of exploiting water from aquifers 400 m below the surface of the West Bank, the waters of part of this 'natural' underground reservoir would be discharged into the Dead Sea as saline water – in which case they would presumably not be defined as a 'natural resource'. Seawater and sewage would not be thought of as water 'resources' unless there existed desalination and wastewater purification technologies, as well as a high social need for water. Conversely, while sweat is not generally considered a water resource – at least not yet! – there is no absolute reason why it could not be, were technological possibility and social necessity to so demand.[57] These points can be generalised. Claims to 'naturalness' always say as much, if not more, about their society of origin as they do about the supposed 'nature' being alluded to, and this goes for 'natural water resources' as for anything else. 'Natural resources' should not be imagined as discretely natural objects.

Contrary to what is implied by Malthusian ecological discourse, nature and society are not ontologically separate. Human encounters with nature are not limited simply to consumption or use, but are also productive, to the extent that there are no natural limits to human economic activity. Rather than merely being consumers of an otherwise discrete nature, human societies are continuously engaged in its purposive transformation and production, human labour being 'a process between man and nature, a process by which man, through his actions, mediates, regulates and controls the metabolism between himself and nature'.[58] Moreover, given that human relations with nature are productive rather than merely consumptive, it so follows that resource thresholds and carrying capacities are functions not of a static and unyielding nature, but of the limited extent of human productive capacities relative to social need. David Harvey puts this point well:

> To declare a state of ecoscarcity is in effect to say that we have not the will, wit, or capacity to change our state of knowledge, our social goals, cultural modes, and technological mixes, or our form of economy, and that we are powerless to modify either our material practices or 'nature' according to human requirements.[59]

To clarify these points, let me put a quick post-structuralist spin on this largely historical materialist account. If we return for a moment to Wolf's account of the surface waters of the Jordan basin, we are presented here with nominalised bodies of water – with stable objects (the Jordan River, the Dead Sea) which cover definable territories, and have quantifiable yields. This account represents water in bodily and territorial terms. Yet water can also be imagined and represented using an wholly antithetical (and much more post-structuralist) spatial vocabulary, one that portrays water as a dynamic fluid that ceaselessly traverses territorial space.[60] Viewed in this way, water is not a stock-like natural resource which is straightforwardly used by human societies, but a restless fluid which journeys continuously between nature and society, and through a cycle that is 'hydro-social' rather than hydrological (from rain clouds to sewage treatment plants and back again). Water in this sense does not exist as a pure, bodily resource, but rather as an impure substance which is recurrently mixing and getting jumbled together with salts, particles and waste. Unlike the orderly and predictable objects depicted in most texts on the Middle East's water problems, water is, from this perspective, stochastic, disorderly and unpredictable.

Seen in this slightly different light, the role of technologies in water management hopefully becomes clear. Water technologies, it may now be said, all function to govern the chaotic movements of a commonly impure fluid: to collect and entrap it (wells, rooftop harvesting systems), to confine it in space (pipelines, water tanks), to purify it (settling tanks, desalination plants), and to control it across time (valves, taps). There are in principle no limits to the application of these technologies, and thus water can in principle be reused and reprocessed *ad infinitum*: the only limits here are in human technological and social capabilities. *Pace* Malthusian accounts – which posit finite stocks of water resources, with natural thresholds and barriers – the truth is rather that flows of water can be recycled as often as economies and technologies allow for, or as often as societies demand. Given this, it can be said that the task of governing and administering water is centrally one of controlling and disciplining recalcitrant and often distant flows of water. This task is productive and creative, and applied as much in the past as it does today. Indeed as one recent textbook astutely observes, 'creating more water than the local environment provides...has been the water engineer's role from time immemorial'.[61]

These points are general ones, applying as much to Dhaka and Baghdad as to Israel and the Palestinian territories. Within the latter two, water is created and made available to local environments first and most obviously by controlling the surface waters of the Jordan basin, and the underground waters of the mountain and coastal aquifers, and then by conveying these waters to distant locales (for instance, from the Sea of Galilee to the Negev desert). Besides this, however, water is also governed – and here we take Israel as an exemplar – through several less obvious and less conventional means. Firstly, Israel is a world leader in wastewater treatment and recycling, such that 70 per cent of its municipal water supplies are reprocessed and reused. Most impressive amongst treatment and reuse systems is the Dan Region Reclamation Project, which recycles the municipal wastewater of the Tel Aviv metropolis, with a population of about one and a half million. Effluent is injected into and then passes through a nearby aquifer basin, the wastewater in the process being subjected to both mechanical and biological treatment. The treated water is then collected through recovery wells, and conveyed to the Negev Desert, where it is used largely for irrigation purposes. Yet despite this current use, the treated water conforms to drinking water standards, and may at some point in the future be used for domestic supply.[62] In principle, then, the water produced from such a scheme could in future be endlessly recycled and reused.

Even more important for Israel and the Middle East, but nonetheless barely noticed, is the import of what Tony Allan refers to as 'virtual water'. During the late 1960s, Israel began to switch the focus of its agricultural production from cereals and other food staples, to the production of high value agricultural crops, and to import food staples from Europe and in particular the US. Besides allowing more water to be used for relatively high value agricultural, industrial and domestic purposes, this policy also in effect meant that Israel was henceforth making use of rain that had fallen in Europe and North America, and is used there for the production of food staples. Allan calculates that the total water and food production needs of the present populations of Israel, the West Bank and Gaza are 7.5 bcmy, which, if correct, would suggest that two-thirds of their total water needs are imported from abroad in barely noticed virtual form.[63] More broadly (since other Middle Eastern states have since followed Israel's lead), Allan contends that 25 per cent of the water needs of the Middle East and North Africa are accessed via imported cereals.[64]

In future, Israel will also no doubt come to rely on water produced through desalination, and on water transported in 'real' (rather than 'virtual') form from neighbouring water-rich states. Israel is currently embarking on construction of its first desalination plant, and is also on the verge of importing water from the Manavgat River in Turkey, the plan being to import 45 mcmy of water across the Mediterranean in 250,000-ton tankers.[65] Many other ideas have been floated as to how Israeli water supplies could be increased, including the conveyance of water across the Mediterranean in giant plastic 'Medusa bags'; or through a 'peace canal' from Turkey, or from the Nile; and the construction of a 'Red–Dead' or 'Med–Dead' Canal, which could at once enable the replenishment of the diminishing Dead Sea, and the generation of hydro-electric power for seawater desalination.[66] While most such schemes are currently viewed as technically, economically or politically unfeasible, they together highlight the point that Israel and the Palestinian territories are not subject to any natural 'water barrier'.

Just as important as this work of governing and conveying water across space, is that of governing water supply and use through time, as a means of adapting to ever-changing and unpredictable patterns of rainfall and resource yield. Water resources, we will recall, do not constitute stable and predictable bodies of water. In Israel's case, it has faced protracted droughts during the early 1990s, and again more recently, yet has nevertheless managed to regulate demand and ensure supplies by overseeing and regulating startling fluctuations in water consumption.[67] Whereas in 1991 (at the height of drought), Israel's agricultural sector used only 875 mcm of water, and total water use was only 1420 mcm, by 1994 agricultural water use had climbed to 1181 mcm, and total water use to 2019 mcm.[68] The Israeli state plainly had the institutional capability to regulate water demand to this degree, and to respond to and cope with extremely unpredictable hydrological patterns. Whether other states and institutions necessarily have such capabilities is an issue that we will return to later.

In conclusion, it can be said that ecological discourse wholly mistakes both the character of the relations between nature and society, and the nature of water. Water resources are not fixed, finite and predictable, but fluid, dynamic and thoroughly unpredictable. This is not to dismiss outright those many analyses of the Middle East's water problems which represent water in stable, bodily form – such representations are generally both necessary and useful for problem-solving, administrative purposes. It is to maintain, however, that water is in reality a mobile fluid, which moves in patterned chaos through natural and social environments, and which poses continual challenges to engineers, technicians and water administrators. Again, this is not to deny that precipitation levels are much lower in some parts of the world than in others, or to deny that the generally low level of rainfall in Israel and the Palestinian territories is one of the major reasons for the existence of a Palestinian water crisis – it would be patently ridiculous to deny this. It is simply to contend that water crises ensue not from natural limits, but instead from a lack of 'will, wit, or capacity' to control patterned flows of water in accordance with social needs.[69]

There are, of course, numerous problems with Malthusian discourse, many of which have not been addressed here: it is simplistic in its assumptions about population, eliding the fact that population growth is often a spur to economic productivity; it is simplistic in its conception of demand, failing to see that, as in the case of water, social needs are always mediated through technologies (the high water demand associated with economic development results from the use of water-consumptive flush toilets and washing machines, not from economic growth *per se*); and it is often very ethico-politically conservative, tending to blame the poor for their over-breeding (but not of course the rich for their high levels of consumption).[70] Perhaps the best that can be said for Malthusian ecological discourse is that its pessimistic rhetoric of natural thresholds can have strong political effects, stimulating political leaders and institutions into action. Against Malthusian ecological discourse, technical discourse provides much the more accurate account of the productive relations between nature and society. It remains to be seen, however, whether technical discourse can really help us to understand the causes of the Palestinian water crisis, and how the technical aspects of this particular crisis are intertwined with its self-evident political dimensions.

TECHNO-POLITICS AND THE POLITICAL ECONOMY OF WATER

From an ideal-typical technical perspective, 'the water problems in most countries stem mainly from inefficient and unsustainable use of water' – from technological and economic inefficiencies, and organisational and policy mismanagement.[71] Seen through a political lens, by contrast, water crises arise primarily from the uneven and inequitable distribution of wealth, power and resources. Clearly, these two types of account are antithetical, not just in their portrayals of the Palestinian water crisis, but also in their very assumptions about the social world and the human condition. Technical discourse is philosophically liberal, being premised on a win-win ontology and assuming the possibility of a felicitous harmony of interests between apparently conflicting goals (such as economic growth and environmental sustainability) and social groups (such as Israelis and Palestinians). It operates under the assumption that, so long as inefficiencies are replaced by efficiencies, and so long as traditional and outdated practices are modernised, then water crises can be overcome to the benefit of all. Political discourse on Middle Eastern water issues is premised, by contrast, on a zero-sum ontology, evincing a social world made up of winners and losers. Unlike technical discourse – which, true to its liberal roots, is simultaneously individualistic and universalistic, imagining a 'common future' shared by undifferentiated individuals – the political worldview has it that social formations are marked by antagonistic social groups, whether states, nations, classes or sexes.[72] Political discourse does generally call for some redistribution of power and resources, and thus typically maintains a fairly strong idea of how progress could, in principle, be brought about; nonetheless it clearly does not submit to the liberal optimism of technical discourse. Technical and political discourse on water crisis are thus sharply opposed: where the one revolves around matters of efficiency, the other lays stress on questions of ownership and control; and where the one is guided by an Enlightenment faith in the possibility of infinite progress, the other bears witness to inequalities and conflicts that are both ongoing, and more than likely to continue.

So conceived, both technical and political discourse are to varying degrees problematic. Technical discourse tends, at bluntest, to ignore power and politics, with the latter of these appearing on stage only as an impediment to reason that must be overcome. Its minimalist explanations are atomistic, depicting the causes of water crisis as a series of discrete absences and inefficiencies within the water sector, rather than within the context of much wider and broader patterns of social relations. Put another way, its explanations are decidedly 'internalist'; they are insensitive, that is, both to the causal relations between problems and issues, as well as – and in this they share much in common with liberal thought as a whole, and with modernisation development theory in particular – to the relational international contexts within which such problems and issues emerge.[73] The World Bank's writings on the Palestinian economy in the West Bank and Gaza illustrate this well (and mirror the Bank's comments, already cited, about the inefficient state of the Palestinian water sector). In its lead report for the international donor community, published just days after the Oslo handshake, the Bank noted of the Palestinian economy that:

> the normal consolidation and rationalization of the industrial sector has not occurred, impeding the realization of economies of scale. The combination of the small size of enterprises, the underdeveloped state of marketing services and the lack of infrastructure and development systems constrains producers to sell directly to customers within a small geographical area... The lack of clear zoning regulations and public land use policy have acted to distort urban/industrial land prices, becoming a barrier to industrial expansion. Finally, business support services and institutions, both public and private, have yet to develop to a stage where they can cater to the needs of a dynamic private sector.[74]

Here, in this not untypical passage, we find the full panoply of technical rhetoric. The West Bank and Gaza economy is not at all 'normal', but is, on the contrary, characterised by a whole series of 'impediments', 'lacks', 'barriers' and 'constraints' which stand in the way of 'development'. These lacks and barriers are of a discrete technical character – things like clear zoning regulations and land use policies which inhibit normal development. They are also, so it seems, internal characteristics that, though their exact origins are far from clear, are nonetheless attributes of a distinct thing called 'the Palestinian economy'. Yet this is in truth very far from being the case. The technical barriers and distortions that the West Bank and Gaza are so characterised by do not follow from a straightforward lack of development, but rather from the specific form of 'dependent development', and in part 'de-development', to which they were made subject under occupation.[75] Under occupation, Palestinian agricultural and industrial development were both intentionally stifled, and the West Bank and Gaza economies came to depend more and more upon poorly paid and non-unionised day labour within Israel. The Palestinians of the West Bank and Gaza also became a captive market for Israeli produce and manufacturing. Thus far from being a discrete economic entity, unfortunately held back by a range of 'constraints' and 'barriers', the Palestinian economy is in reality a dependent and strongly incorporated part of the larger Israeli economy. The very idea of a 'Palestinian economy' is to this extent the largely fictitious construct of economic planning and development discourse.[76] Likewise the very idea of a 'Palestinian water crisis' needs to be understood as signifying not an internally generated crisis, but one that, while located in the West Bank and Gaza, nonetheless has to be explained as the product of a set of structured political economic relations with Israel and the world system at large. Technical discourse tends to hide, elide or simply misunderstand such matters. In this respect political discourse provides much the better basis for understanding both water crisis in general, and the Palestinian water crisis in particular.

This is not to say though that political discourse is without its problems. The first of these is that while strongly technical accounts typically say little of the broader political and economic contexts within which technical problems arise and are reproduced, more politically oriented accounts tend equally to understate the parts played (or not played) by techniques and technologies in creating inequalities. Take, for instance, the account offered by Jad Isaac. Isaac, as we have already seen, defines the Palestinian water crisis in primarily political terms, as the consequence of an uneven and inequitable distribution of the region's waters. Expanding on this, he notes that

> Palestinians are prevented from fully utilizing the West Bank's underground water resources. Permission for well-drilling must be obtained from the military authorities; permits have been granted for only 23 wells since 1967, only three of these being for agricultural use... Rigorous water quotas are imposed on Palestinians, supply is often restricted leaving communities without water for considerable periods, and excess pumping is punished by heavy fines.[77]

Isaac, it should be said, is wholly correct in pointing to the many constraints imposed by the Israeli occupation authorities upon Palestinian water use (I will detail some of these in chapter 3). Nonetheless, it is worth noting that the Palestinians, as represented by Isaac, are subject to what, following Foucault, may be thought of as an essentially 'negative' form of power, a mode of power 'which lays down the law, which prohibits, which refuses, and which has a whole range of negative effects: exclusion, rejection, denial, obstruction, occultation'.[78] Power is concentrated, according to this model, in the hands of the state, which acts oppressively and negatively to constrain social activity. The Palestinians, as Isaac depicts them, are subject to a power that is possessed by the Israeli state, and that acts oppressively – through a 'negative' array of 'preventions', 'impositions', 'restrictions' and 'punishments' – to deny the Palestinians control and use of their rightful water supplies. In large part, this is indeed an accurate portrayal. Yet if we are to believe Foucault, modern governmentalised power is neither simply concentrated in the hands of a powerful sovereign, nor primarily repressive in its form. Foucault contended to the contrary that 'the interdiction, the refusal, the prohibition, far from being essential forms of power, are only its limits, power in its frustrated or extreme forms. The relations of power are, above all, productive.'[79] In light of this he argued that we should, metaphorically speaking, 'cut off the King's head', and focus our analyses not on questions of 'sovereignty-oppression', or on the 'solid domination' that one group exercises over another, but rather on the micro-scale technical apparatuses, procedures and mechanisms through which power is practised, articulated and consolidated.[80] We should focus, suggests Foucault, not on the 'what' or the 'why', but instead on the 'how of power'.[81]

Now admittedly, Foucault does somewhat overstate the dispersed character of modern power, paying too little attention to those forms and practices of power that exist at its most negative and repressive limits. That said, Foucault's emphasis on the 'micro-physics of power' calls welcome attention to the diverse ways in which techniques and technologies are, at a microscopic level, regularly embroiled in politics.[82] The pertinence of this is that the Palestinian water crisis should not be thought of simply as the product of 'policy', exercised at a sovereign centre by the powerful Israeli state, but also as the product of those innumerable technologies and techniques which are both part and parcel of the Israeli state's power, and central to its capacity to put its discriminatory policies into practice. Scientific knowledge, technical expertise and engineering systems all play a role in mediating Israeli policy, and in giving the Israeli state the power that it has. Political discourse tends to focus on policy, seeing the Palestinian water crisis as a function of policy decisions, where the key actors are those politicians and generals who formulate policy. But such an approach is, to adopt Langdon Winner's felicitous phrase, 'technologically somnambulist'.[83] What is needed in addition to analyses of political actors and their policies is attention to the diversity of technical work done by those hydrologists, water planners, water managers and civil engineers who work within the Israeli–Palestinian water arena, and without whose input the Palestinian water crisis would have a very different character from that which it has today.

A second problem with political discourse, certainly as it features in the Israeli–Palestinian arena, is that it concentrates above all on one particular level or plane of political conflict. In keeping with its broadly problem-solving orientation, political discourse on the Palestinian water crisis says nothing of the global political economic contexts within which that crisis has emerged and been reproduced. More strikingly still, such political discourse tends to say very little about water supply inequalities and conflicts within Israel and the Territories. Political discourse on

Israeli and Palestinian water issues instead focuses almost exclusively on the water conflict between the two peoples, Israeli and Palestinian, these being standardly depicted as wholly undifferentiated groups. West Bank Palestinians are typically presented as facing their water crisis *en masse*. What such depictions elide, however, is the fact that there are very large differences in water supply, and in control of water resources, between and even within Palestinian communities. Within the existing literature, there is very little discussion of these important local inequalities and conflicts – in large part, it seems, because of the strongly nationalist tenor of public discourse within the Israeli–Palestinian arena.[84] None of this is to deny the immense average differences in water supply that are received by Israelis and Palestinians. It is to suggest, however, that accounts of this central axis of difference and discrimination need to be complemented with analyses of water supply conflicts occurring at various sub-national sites and scales. Politics, after all, is not just the preserve of the state.

Quite apart from my use of it here, the distinction between technical and political considerations is one that regularly arises within expert discourse and rhetoric. Water experts commonly use it to define their own claims as impartial and unbiased, and to define those of their leaders as driven by self-interest and the demands of political survival ('Arafat is not a technical person; he's a complete political animal,' observed one expert to me – he was clearly glad to be a person). Israeli experts and policy-makers invoke it to argue that talks should centre on technical matters such as data gathering, supply enhancement and water management, while Palestinian officials do likewise in arguing that negotiations should centre on political questions of distribution, ownership and rights.[85] The World Bank uses it to define itself as an economic institution, one that produces technical reports and provides technical advice without entering into the dirty world of politics.[86] Besides this, the distinction is also at least implicit in a great deal of commentary on Middle Eastern water issues. Aaron Wolf, for instance, identifies two distinct problems in the Jordan basin, one a 'water crisis' in the imbalance of supply and demand, the other a 'water conflict' between politically hostile riparians.[87] Miriam Lowi invokes it in distinguishing, as we will see at greater length in the next chapter, between the 'high politics' of war and diplomacy and the 'low politics' of inter-state technical co-operation.[88] Alwyn Rouyer, meanwhile, depicts the ongoing Israeli–Palestinian water conflict as a function of 'Israel's internal politics' and 'founding myths', his main suggestion being that more rational economic and technical management would pave the way for the conflict's resolution.[89] Right across the field of discourse on Middle Eastern water issues, the technical-political distinction is invoked with striking regularity.

Yet the distinction is problematic on at least two grounds. In the first place, and most obviously, it is frequently used rhetorically to justify certain types of action and inaction, while de-legitimising and excluding others (as Bruce Rich says of the World Bank, for instance, the 'Bank's definition of "politics" and "political influence" appears to come straight out of *Alice in Wonderland*: whatever it does is by definition not political because it says so').[90] More importantly in the present context, the constant invocation of the technical-political distinction within analyses of and commentaries on Middle Eastern water issues tends to generate an elision of questions of the causal relations and connections between techniques and politics. By contrast with this typically problem-solving way of describing and intervening in Middle Eastern water problems, the task of explanation demands analyses that focus instead primarily on the relations between the technical and the political dimensions of water crisis. This implies several different things. At a macro-level, it implies the need not for separate analyses of the politics and the economics of Middle Eastern water issues, but instead for singular analyses of the political economy of water – of the specific character of the relations between state and economy, and between distinct social groupings, within Middle Eastern states; of the place of these states, economies and social groupings within the larger capitalist world system; and of the particular ways in which structured patterns of political economic interaction end up producing and reproducing local water problems. Equally at the micro-level, what is needed are not separate descriptions of technical inefficiencies and political impediments – though these are of course important – but in addition, explorations of the multiple ways in which techniques and technologies are used to harness and produce water resources in the service of state authorities; of the various ways in which techniques and technologies are complicit in Israeli–Palestinian water relations and the Palestinian water crisis; and of the local supply inequalities and conflicts whose causes are always a complex *mélange* of technical and political factors.

This conclusion relates not just to the Israeli–Palestinian arena, it should be said, but to water crises in general. Malthusian ecological discourse misapprehends both human-nature relations, and the ways in which nature is continually produced through human labour. Technical discourse errs in its insensitivity to structural political economic contexts. Political discourse, while correct in its social and non-atomistic ontology, and in its emphasis on the prevalence of differences, inequalities and conflict, is nonetheless often insensitive to the technological 'how' of water conflicts, and is unfortunately often couched in purely nationalist and state-centric terms. Water crises in general are not problems either of Malthusian limits, nor simply of either technical deficiencies or group conflicts, but are instead always political economic problems which have both macro-scale and micro-level 'techno-political' causes. This applies with equal weight, though in various different ways, to all of the individual cases discussed at the outset of this chapter. And if the Israeli–Palestinian case might seem more purely 'political' in its causes than the crises in either Bangladesh or Malawi, then hopefully the analyses to follow will suggest to the contrary.

CHAPTER 2

Explaining Water Conflict in the Jordan Basin: Beyond the Realist–Liberal Orthodoxy

Over the past twenty years or so, Western water experts have been engaged in a protracted argument over the likelihood of water wars breaking out in the Middle East. Two camps have been clearly discernible. On the one hand have been 'the pessimists', typically contending that water shortages have already been an important factor in causing, or at least contributing to, inter-state conflicts in the Middle East, and maintaining that these shortages will become ever more weighty sources of instability in the future. Pitched against them have been 'the optimists', who tend to deny that water shortfalls have caused any of the Middle East's violent twentieth-century conflicts, and tend also to doubt whether water wars lie on the horizon. In this debate between pessimists and optimists, the positions are clearly marked, the arguments and counter-arguments replayed with stubborn regularity.

Often coinciding and overlapping with this water wars debate has been a second, generally more reflective one, about the causes of and reasons for inter-state conflict and co-operation over the Middle East's water resources. This debate, with its no less entrenched positions, has tended to pit 'realist' scholars with their emphasis on the ceaseless inevitability of power struggle, against 'liberal' writers, who tend to evince at least the possibility of reasoned progress in the international arena. As with the water wars debate, viewpoints here are starkly divided, and tend to get repeated and revisited with a frequency that shows no signs of abating.

For understandable reasons, realists tend to be more pessimistic about the Middle East's hydro-political future, liberals quite the reverse. For instance, in a recent rebuttal of the hydro-pessimists and their realist assumptions, Mostafa Dolatyar and Tim Gray develop an argument that is quintessentially liberal as well as optimistic. Their contentions are threefold: first, that water scarcity has not been a fundamental cause of any of the twentieth century's Middle Eastern wars; second, that deepening water scarcities will not engender violent conflicts in the future, since 'water is too precious to risk by going to war'; and third, that 'moves towards settlement of water disputes could promote efforts at achieving wider peace objectives'.[1] While expressing doubts over some liberal arguments, Dolatyar and Gray nonetheless continue to sing from a paradigmatically liberal song-sheet. The twentieth century may have passed, but the fault line between pessimistic and optimistic water experts remains as wide as ever.

It should be readily apparent from the previous chapter that I have little time for Malthusian reasoning about water wars. Malthusians operate with a mistakenly reified understanding of nature, and thus also of natural limits, the consequence of this being that they tend to see famines, ecological crises and in turn wars as the likely and even inevitable products of population growth and development. Many water wars arguments are founded, explicitly or otherwise, on just such Malthusian premises. To assess, however, whether all pessimist and indeed realist arguments are thereby mistaken, we need to consider these arguments head on. This chapter attempts to do precisely that: to consider the historical and logical plausibility of the pessimists' water wars arguments; and more broadly, to move on from the previous chapter's consideration of the causes of water crises to a more international relations-oriented investigation of the causes of conflict and co-operation over the Jordan basin's trans-boundary water resources.

As will shortly become apparent, I view the claims of hydro-pessimists as lying somewhere between overblown rhetoric and Orientalist fantasy. I am also sceptical, however, of both liberal-optimist and realist arguments about Middle Eastern water conflicts, in the first case because of their usually glaring political naïvety, in the second because of their insensitivity to historical change. What is needed instead, I suggest, are more fully historical, social, political and political-economic analyses of the causes of inter-state water conflict, ones that move beyond the mythical idea of 'water wars', as well as beyond the rather stale terms of the orthodox realist-liberal water politics 'debate'.

I begin by developing three subsequent critiques – first of the pessimistic claims of the water wars brigade, and thereafter of representative 'liberal-optimist' and 'realist' takes on the Middle East's water politics. On the strength of these critiques, I develop in the fourth section a Marxist-informed account of the historical foundations of Israeli water policy. It was argued in the previous chapter that, at a macro-level, analyses need to focus above all on the political economic causes of water crisis, rather than on discrete spheres of activity labelled 'political' and 'economic'. The latter part of this chapter endeavours to do precisely that, and to provide through so doing a necessary historical backdrop for the more detailed and micro-scale analyses developed in subsequent chapters.

THE SPECTRE OF WATER WARS

In a much-cited text published in 1988, Joyce Starr and Daniel Stoll of the Center for Strategic and International Studies in Washington DC observed that:

> The Middle East stands at the precipice of another major natural resource crisis. Before the twenty-first century, the struggle over limited and threatened water resources could sunder already fragile ties among regional states and lead to unprecedented upheaval within the area.[2]

And they are far from being the only ones to have issued such alarming forecasts. In May 1990, the then Egyptian Foreign Minister (and later UN Secretary General) Boutros Boutros Ghali gave advance notice that 'the next war in the Middle East will be over water, not politics,' while in 1995, the World Bank's Ismail Serageldin ruminated that 'many of the wars of this century were about oil, but wars of the next century will be over water'.[3] For Joyce Starr, resolving water conflicts in the Middle East is now no less than the 'key to world survival'.[4] At least the last two of these prophecies have not yet been proved wrong.[5]

Others have developed their water wars arguments along historical lines, with the favourite exemplar here being the 1967 war between Israel and the Arab states of Egypt, Jordan and Syria. John Bulloch and Adel Darwish contend for instance that the '1967 war – the Six Day War – was caused largely by competition for the waters of the River Jordan'.[6] Thomas Naff and Ruth Matson observe that the 'increase in water-related Arab–Israeli hostility was a major factor leading to the 1967 June War'.[7] John Cooley, meanwhile, has argued that the 'constant struggle for the waters of the Jordan, Litani, Orontes, Yarmuk and other life-giving Middle East rivers ... was a principal cause of the 1967

Arab–Israeli war'.[8] Such claims are commonplace within the Middle East water politics literature.

For all their apparent seriousness, however, the debilitating problem with these water wars readings of the 1967 war is their total variance with informed histories of the conflict. Not only do the likes of John Cooley make basic empirical errors in their accounts of this 'water war' (contrary to what Cooley says, neither the Litani nor the Orontes were shared by the protagonists in the 1967 war); they also completely elide its central political causes. As informed histories make clear (though emphases of course vary), Israel launched 'pre-emptive' attacks on Egypt, Jordan and Syria not because of water resource disputes, but in an attempt to shatter Nasser's Arab Nationalist prestige, to enhance the country's strategic depth, and to fulfil longstanding Zionist territorial ambitions; more broadly the war was the product of Cold War bipolarity, of poor intelligence information, of the political rivalries between Arab states, and of the political insecurities of Israel's Eshkol-led government.[9] It can quite reasonably be argued that Israel's invasion and occupation of the Golan Heights towards the end of the war (which took place despite the fact that Syria had not been a direct party to the war, and had already accepted the UN's call for a cease-fire) was in part inspired by hydro-strategic purposes. Nonetheless, in general it can be said that 'high political' issues are rarely addressed in hydro-pessimist accounts of the Middle East's supposed water wars, with the consequence that they end up grossly exaggerating the international political significance of water scarcities.

This historical evidence aside, there is a certain apolitical incoherence to the oft-made prognoses of water wars on the horizon. The point is well illustrated by Boutros Ghali's portrayal of water, not politics leading the Middle East into war, as well as by Joyce Starr's frankly incredible claim that '[w]ater security will soon rank with military security in the war rooms of defense ministries,' and by the regular assertions that 'water has become a commodity as important as oil,' and that 'water, not oil, will become the dominant subject of conflict for the Middle East by the year 2000'.[10] *Pace* Boutros Ghali, wars are always a product of 'politics' – whether conceived in terms of political interests and strategies, political structures and values, or of political motivations and decisions – and this applies as much to water as to any other actual or potential cause of war. It is impossible to imagine a war that did not involve politics. Against Starr, while environmental security issues have undoubtedly become of greater interest to NATO, the Pentagon and so on in recent years, there is nevertheless no direct parallel between water security and military security: water might, just conceivably, contribute to the outbreak of future wars, but the central concerns of defence ministries and military institutions will continue to be military ones. Most saliently, against those who prophesy that water will become the strife-torn blue gold of the twenty-first century, such claims wholly elide the high politics and political economy of resource conflicts. Oil has been such a source of conflict in the Middle East precisely because of its importance to political and economic elites, providing local regimes with much needed rents for development, patronage, and the build-up of military arsenals, while at the same time being the primary lubricant of the contemporary capitalist world system. Iraq invaded Kuwait in 1990 primarily because the latter's bountiful supply of oil served as an attractive source of revenue, at a time when the Iraqi regime was struggling to repay its debts and reconstruct the country's infrastructure in the wake of its eight-year war with Iran, while for its part, the US-led coalition responded as it did in large part because of the perceived threat to international oil supplies, amassing an incredible 500,000 troops along the border with Saudi Arabia. It is hard to imagine water ever being of sufficient value as an economic commodity to prompt military action on this scale, either by Middle Eastern regimes or by the international community. Water is no doubt a biological and ecological *sine qua non*, and water shortages doubtless also severely impact on the everyday lives of Middle Eastern peoples. Unlike access to oil, however, water shortages are generally (though with some important individual and historically specific exceptions) of only marginal concern to Middle Eastern elites, let alone to international ones.

The spectre of water wars, in sum, is little more than that. The rhetoric of water wars doubtless serves some important purposes, since scaremongering by political leaders and experts is an easy way of prompting attention, intervention, and development assistance packages (as Tony Allan puts it, 'the pessimists are wrong but their pessimism is a very useful political tool').[11] Rhetoric aside, though, the idea of water wars also seems to carry with it traces of simplistic 'Orientalist' stereotypes about the Middle East.[12] As an idea, it no doubt appeals to Western images of barren deserts and dangerous Arab dictators – it is no co-incidence that a book such as Bulloch and Darwish's *Water Wars: Coming Conflicts in the Middle East* portrays Saddam Hussein and Colonel Qaddafi on the cover – as well as to what has been called the prevailing 'myth of instability' of the Middle East.[13] Whether this is the case or not, forecasts of coming water conflicts tend to combine poor history, apolitical reasoning and overblown rhetoric – and thus should not be taken too much at face value. To understand the international politics of water in the Middle East, we really need to look elsewhere.

WISHFUL THINKING ON WATER AND PEACE

Dolatyar and Gray's recent foray into the water wars debate is as good a place to start as any, since it develops a case that is both paradigmatically liberal-optimist in argument, and explicitly and rightly critical of the notion of water wars.[14] This, at least, is how things appear at first glance. Scrutinised more thoroughly, however, it becomes clear that Dolatyar and Gray's argument shares a great deal in common with the water wars arguments discussed above. To explain how this can be, we need first to know something of the broad contours of liberal internationalist thought, and of its long-time antithesis-cum-alter ego, realist international relations.

With E.H. Carr, Hans Morgenthau and Kenneth Waltz as its canonical twentieth-century figures, realist international relations depicts international politics as a realm of inescapable uncertainty and danger, where state action is governed by a constant struggle for power and security.[15] For realists, the foundational truth of international politics is the absence of world government, of a supra-national Leviathan that might impose order on states. Hence from a realist perspective, the international arena is no less than a negative mirror image of domestic political life: domestic politics is marked by the sovereign presence of government, international politics by its anarchic non-existence; the one bears witness to law, regulation and at least potential order, the other to no law, and no order, other than that prescribed by power; the one is thus characterised by the possibility of progressive development, the other by constancy and repetition, an eternal return of self-interested power plays. From a realist perspective, states are the main and indeed the only really important actors. In the absence of world government, states act purely out of their national interest, this interest being to amass as much power and security as is possible – by constantly enhancing military, economic, diplomatic and propaganda capabilities, and by forming useful, though certainly not trust-based, tactical interstate alliances. For realist international theory, the national interest is 'defined as power'.[16] Given this, state action in the international arena is driven not by ideas, values or ideological preferences – by foreign policy preferences for fellow democratic or communist or Islamist states – but by the dictates of necessity and self-help within a hostile environment. We may not like it, but that for realists, is the way international politics is, the way it has always been, and the way it will continue to be. The 'texture of international life has remained impressively, or depressingly, uniform even while profound changes were taking place in the composition of states,' observes Waltz.[17] This is the timeless reality of international politics.

From a liberal internationalist perspective, by contrast, progress is possible within the world of international politics. Sovereign government is absent, yes, but that does not mean that the realm of the international is totally lawless. State action is driven not only by the dictates of power, but also by normative and value commitments – one strand of liberal internationalist thought argues, for instance, that liberal democracies simply do not go to war with one another.[18] Moreover, *pace* realism, states are increasingly tied together in complex webs of economic interdependence, which lead to a flourishing of shared interests and militate, for quite rational economic reasons, against the outbreak of inter-state war.[19] States increasingly share their power with international institutions, corporations, non-governmental organisations and so on, being no longer the sole locus of political agency and power. Thus while in the past the international system was largely dominated by the play of self-interested power, today it is increasingly witnessing the development and acceptance of consensually shared rules, norms and values. Such, at least, is international relations as seen through liberal-tinted glasses.

While realists tend to see disorder and conflict as the normal state of affairs within international politics, liberals typically emphasise instead the orderliness and co-operative character of inter-state activity. While for realists, 'states are predisposed towards conflict and competition, and… often fail to cooperate even when they have common interests,' for liberals precisely the opposite is true.[20] Realists and liberals also offer starkly contrasting types of explanation for conflict and co-operation. From a realist perspective, while conflict is undoubtedly the more natural condition, co-operation can and does of course occur – but because of, and not in spite of, prevailing power relations. Co-operation occurs either through the dominance of a single hegemonic actor (such is the central tenet of 'hegemonic stability theory') or alternatively, and realists vary in their emphases here, when co-operation 'roughly maintains preco-operation balances of capabilities'.[21] Conversely, liberals of course recognise the existence of war and conflict, as well as co-operation. Liberals tend to contend, though, that conflicts arise not from imbalances in the distribution of power, but rather from misperceptions, inadequate knowledge, and poorly designed or undemocratic administrative and political structures. Seen in this light, co-operation flourishes when the necessary technical measures are taken – that is, when international institutions, or 'regimes', are appropriately designed, when governments are democratic and accountable, and when decision-making power is transferred from politicians to expert-led 'epistemic communities'.[22] Co-operation may also expand through what 'functionalist' liberals such as Mitrany refer to as a 'spill-over' effect, when functional co-operation between states over 'low-political' issues (welfare and economic policy, for instance) fosters greater understanding and in turn co-operation in more 'high political' areas of policy-making (that is, in defence and security policy), leading to ever-greater interdependence and co-operation.[23] Thus what distinguishes realism and liberalism, in the final analysis, is not just their respective descriptive privileging conflict and co-operation, but their starkly contrasting types of explanation of these contrasting empirical phenomena.

In terms of its descriptive privileging, Dolatyar and Gray's argument is most certainly liberal, being premised on the claim that 'although water has sometimes provoked tension and dispute in the Middle East, it has much more often promoted co-existence and cooperation'.[24] Quite apart from this, though, Dolatyar and Gray also adopt a quintessentially liberal explanatory perspective. Two such liberal arguments stand out. First, on the question of whether deepening water scarcities will inspire violent conflicts in the Middle East, they submit that water shortages foster interdependencies: '[W]ater scarcity creates a mutual hostage situation between riparian states of shared river basins, and this leads such states to avoid conflict by pursuing mutually beneficial solutions to the problems. In other words, water is too vital a resource to be put at risk by war.'[25] Second, on the question of whether water diplomacy can 'spill over' in functionalist fashion to assist peacemaking more broadly, Dolatyar and Gray answer, 'quite possibly, yes... moves towards settlement of water disputes could promote efforts at achieving wider peace objectives'.[26] Both of these arguments are paradigmatically liberal. Both of them also testify, however, to wishful thinking in the extreme.

Let us begin with the second of these claims, namely that co-operation over water issues might perhaps spill over into the realm of high politics, paving the way for ever-greater co-operation between otherwise hostile enemies. Dolatyar and Gray assert not only that this is possible, but that such a 'spill-over effect' has already occurred, 'water scarcity,' they assert, 'has invariably been a platform for cooperation in the region'.[27] The problem here, though, is that Dolatyar and Gray offer no meaningful evidence to support their assertion. They marshal plenty of upbeat words about the effects that water co-operation might have – Daniel Hillel's claim that 'water *can* catalyse and lubricate the peace process…and soften the transition to regional co-operation,' Aaron Wolf's equally liberal contention that water *could* 'induce ever-increasing co-operation…between otherwise hostile riparians, in essence "leading" peace talks' – but they provide no grounds for thinking that this is any more than idle speculation.[28] Indeed they offer only one piece of supporting evidence, that of water co-operation between Bahrain and Saudi Arabia, and even this provides no reason to believe that water scarcity has been a 'catalyst for co-operation', since co-operation between these two states is long-established, and maintained courtesy of Saudi hegemony.[29] There is in fact very little record of water co-operation developing in the face of inter-state enmity. The one notable case of lasting co-operation developing within a context of high political hostility – that of Indo–Pakistani co-operation over the Indus basin, which has been governed since 1960 by the terms of the Indus Waters Treaty in spite of the continuing low- and intermittently high-level military conflict between the two states – lends absolutely no support to functionalist reasoning, this treaty having had no impact at all in stimulating any broader *rapprochement* between India and Pakistan.[30] Against Dolatyar and Gray's functionalism, there is no good reason to believe either that water co-operation has ever been 'a stimulus to international peace making,' or that it might at some future date become one.[31]

As for Dolatyar and Gray's other argument – that water scarcity is unlikely to give rise to wars – this can be dealt with briefly, since the question of water wars has already been addressed. Dolatyar and Gray's reasoning here is twofold: first that water shortages create a 'mutual vulnerability' between riparian states, and second (by way of a 'deeper explanation') that water is simply 'too precious to fight over', the result of these twin factors being that states are inexorably driven towards co-operation.[32] However, both of these arguments are misguided. In the first case, upstream and downstream states to a shared water resource are far from being in a 'mutual hostage situation' since, hydro-politically at least, downstream states are in much the more vulnerable position (upstream states can 'turn off the tap', downstream ones cannot). More strikingly, the claim that water is 'too precious to risk by going to war' is simply naïve in the extreme. As we have already seen, oil has been a cause of conflict in the Middle East precisely because of its preciousness to local and international elites. Preciousness is no barrier to conflict. It is true that water has not been a direct cause of inter-state war in the Middle East, but this is not because of its great political and political economic significance. Quite the reverse in fact: water is generally of relatively marginal importance to the political economy of the Middle East, and it is for this reason, above all, that it has not engendered any water wars.

As will hopefully be apparent by now, Dolatyar and Gray's representative liberal-optimist arguments suffer from thoroughly apolitical reasoning. Saudi–Bahraini water co-operation is portrayed as a 'catalyst for co-operation' – but this assessment is only reached by ignoring the bigger picture of Saudi–Bahraini relations. Water is presented as being 'too precious' to breed conflict – but this conclusion is founded on politically naïve ideas about the relations between preciousness and war, as well as on a mistaken belief in the value of water. What is particularly striking here is that, for all Dolatyar and Gray's criticisms of water wars arguments, they in fact share with them an elision of the high political and political economic contexts of water conflict and co-operation. In this regard it is liberal-optimists rather than realists who hold the most in common with the water wars brigade. Neither can provide a convincing or even plausible explanation for patterns of conflict and co-operation over the Middle East's waters. What then of our third candidate, realist international theory?

UNREALISTIC REALISM

The most theoretically engaged and also most explicitly realist work to have been written thus far on Middle Eastern water politics is Miriam Lowi's *Water and Power*. It is an impressive work, presenting both a keenly researched analysis of the history of conflict and co-operation over the water resources of the Jordan basin since the establishment of the state of Israel, and a tightly argued case study in realist international theory. Empirically, Lowi's account comes in three parts. It starts out by focusing on a failed attempt at co-operation, describing US-mediated efforts during the mid-1950s to construct a basin-wide institutional structure for the management of the Jordan River, an institution to which Israel, Jordan, Syria and Lebanon would all, it was hoped, have been party. It then turns to a period of conflict, providing an account of the 'Jordan waters crisis' of 1964–67, and of the part played by this hydro-political dispute in contributing to the Arab–Israeli War of 1967. Finally, Lowi turns to the period since 1967, focusing on conflict and co-operation within the southern segment of the Jordan basin, between Israel and Jordan, and between Israel and the West Bank's occupied Palestinian population.

Besides mapping out this history of inter-state water relations, *Water and Power* also provides a theoretical analysis of the contending merits of realist and liberal approaches to international relations, Lowi's broad aim being to assess, using her wealth of historical-empirical evidence, which of these two perspectives best explains when and why states co-operate over 'low political' matters such as water management, and when and why they instead engage in conflict. Lowi's conclusions are clear. On the strength of her historical analysis of Jordan basin water politics – and with additional empirical support derived from brief overviews of the hydro-politics of the Euphrates, Indus and Nile rivers – Lowi submits quite simply that 'the realist critics of functionalism are correct':

> states that are adversaries in the 'high politics' of war and diplomacy do not allow extensive collaboration in the sphere of 'low politics', centred around economic and welfare issues. In fact, the spillover effect runs in the opposite direction to that suggested by Mitrany: economic and welfare collaboration is retarded by 'high politics' conflicts between states.[33]

This conclusion leads Lowi to assert that the United States' endeavours during the 1950s and 1970s to promote co-operation over water in the Jordan basin without addressing the 'high politics' of the Arab–Israeli dispute were 'misguided'.[34] Against functionalist and other liberal arguments, Lowi maintains that co-operation over international river basins occurs through the combined effect of just two factors, 'relative power' and 'resource need'.[35] On relative power, Lowi asserts that 'outcomes' in international politics 'reflect the distribution of powers', and more specifically, that co-operation 'in international river basins is brought about by hegemonic powers'.[36] As for resource need, Lowi suggests the following as a general rule:

> The state which is the furthest upstream and hence, in the most favorable geographic position, will have no obvious incentive to cooperate. Being at the source of the river, it can utilize as much of the water as it chooses unilaterally, irrespective of downstream needs. It will not cooperate unless coerced to do so... In contrast, downstream states... irrespective of their relative power resources,

> will seek a cooperative solution because, given their inferior riparian position, they are needier than and, at least in theory, at the mercy of those upstream.[37]

Much of this is persuasive. Lowi is right to stress, *contra* liberal functionalism, the extent to which 'low political' co-operation tends to be retarded by 'high political' conflict. Moreover, *Water and Power* is judicious in its assessment of the place of water in the Arab–Israeli conflict, noting of the 1967 war, for instance, that the Jordan waters question merely 'exacerbated the tensions ... that eventually led to military confrontation,' and that the riparian dispute can best be regarded as 'a manifestation, or dimension, of the Arab–Israeli conflict'.[38] Lowi's account, while realist in its conceptualisation, certainly does not endorse a hydro-pessimist reading of the Middle East's recent history – it is far too politically sensitive for that. Power and interests are clearly key driving forces behind state action in international river basins, and in these respects *Water and Power* is both cogent and convincing. Nonetheless, Lowi's argument does fall short in at least two regards.

A first problem lies in Lowi's realist commitment to the idea of the 'national interest'. Just as realists in general assume the existence of objective national interests, defined in terms of the maximisation of power and security, so too Lowi assumes the existence of objective 'resource needs' – that is, objective state interests in maximising access to scarce water resources, and in engaging in co-operation or conflict in accordance with this imperative. In this schema, patterns of conflict and co-operation are determined solely by such national interest and power considerations. This, however, is implausible, as can be clearly illustrated through a brief review of Israeli–Jordanian relations.

In her historical account, Lowi quite rightly observes that despite the failure of US-led attempts to construct a formal basin-wide water-sharing regime during the 1950s, Israel and Jordan largely abided, from the late 1950s onwards, by a set of tacitly agreed water quotas, and maintained a high degree of informal technical contact over water issues. During this period, neither Israel nor Jordan officially recognised one another, and indeed no Arab state had yet recognised, or reached a peace agreement, with Israel. It would appear, then, that what we had here was the emergence of a limited *modus vivendi* over the low political issue of water supply, despite the continuing discord within the high politics of the Arab–Israeli conflict.[39] How should this unofficial co-operation over water issues be explained? For Lowi, this Israeli–Jordanian *modus vivendi* is nothing less than the inevitable result of patterns of resource need and power relations across the Jordan basin. Prior to 1967, both

Israel and Jordan were downstream of Syria, and hence potentially vulnerable to Syrian over-exploitation of the Jordan. However, following its capture of the Golan Heights in 1967, Israel became an upstream state; Jordan, on the other hand, remained a downstream state, occupying a markedly inferior riparian position. Moreover, while Israel has been the regional hegemon at least since 1967, Jordan has been economically weak and politically fractured. 'Israel's power resources are far superior to those of Jordan,' notes Lowi, and hence because of 'both the power constraints upon the Kingdom [of Jordan] and its critical need for access to scarce water resources, it has tried to maintain, in the aftermath of the 1967 defeat, a non-confrontational relationship with Israel'.[40] In keeping with her broad thesis, Lowi thus explains Israeli–Jordanian co-operation in terms of the twin factors of resource need and the distribution of power.

It is without doubt the case that Israel is a more powerful state than Jordan, and equally true that Jordan occupies a deeply vulnerable riparian position within the Jordan basin. Nevertheless, these two factors are not in themselves sufficient to explain Israeli–Jordanian co-operation over the Jordan River. Against realism, national interests and state projects and policies always emerge within the context of particular domestic struggles, structures and relations, and Jordan is no exception. Ever since gaining its independence in 1946, Jordanian national interests have been defined in thoroughly conservative terms by its British-installed Hashemite monarchy. Thus during the 1950s, while Egypt, Syria and Iraq were all witness to the overthrow of pro-Western regimes, Jordan remained committed to, and received military, diplomatic and economic support from, Britain and somewhat later the United States. Well before the signing of a formal peace treaty in 1994, there was a long history of covert co-operation between the Zionists, later Israel, and the Hashemite monarchy. With a shared interest in containing the development of Palestinian nationalism, the pre-state Zionists and Hashemites entered during the British Mandate era into what Avi Shlaim has referred to as an 'adversary partnership'.[41] During the first Arab–Israeli War of 1948–49, Israeli and Jordanian troops barely engaged, seemingly on the basis of a tacit agreement between the two over the fate of the West Bank – which, as a result, was left well alone during the war by Israeli forces, and was instead occupied and then annexed by the Hashemite Kingdom.[42] Co-operation between Israel and Jordan continued from then onwards, on a 'high political' as well as a 'technical' level, and over a wide range of issues, not just water: agricultural, industrial, health, transport, tourism, air traffic control and mosquito eradication issues, to name but a few, were all, at one time or another, a focus of discreet Israeli–Jordanian co-operation.[43] Israeli–Jordanian relations had their ups and downs, of course, in part reflecting changes internal to the two states (co-operation deteriorated, for instance, in 1977, following the election of Menachem Begin and the Likud).[44] Nonetheless, throughout the Cold War period there was a consistent level of informal co-operation between Israel and the Hashemites.

It is within these various domestic, international and long-historical contexts that the so-called water-related 'picnic table summits' between Israel and Jordan – which occurred two to three times a year at the confluence of the Jordan and Yarmouk Rivers, right up until 1994 – have to be understood.[45] US economic aid for Jordanian water projects was made conditional on Jordanian adherence to water quotas with Israel: thus by locating itself within the Western orbit during the Cold War, Jordan inevitably found itself being pushed towards co-operation with Israel.[46] Israeli–Jordanian water relations have doubtless been strongly shaped by questions of state power and need. But they have also been shaped by historically specific struggles, structures and forces, ones that Lowi and realist theory in general, in their exceedingly narrow focus on state capabilities and state needs, typically overlook. Jordanian water policy would have been totally different had Jordan undergone, say, an Arab nationalist revolution. Jordan co-operated with Israel on water issues not because of an objectively real 'national interest', but because of the longstanding projects, strategies and sub-national interests of the ruling Hashemite monarchy, and the inter-state relations that developed on the strength of these within the context of the Cold War international system.

To be fair Lowi is aware of some of this, alluding during the course of her historical survey to the 'Jordanian tradition of accommodation with Israel'.[47] Lowi's empirical analysis is sensitive to the complexities of 'history, culture and ideology' and, unlike so much contemporary realism (which generally goes under the name of 'neo-realism') thankfully does not seek to utilise rational choice and game theoretical models of political behaviour.[48] Lowi even observes in conclusion that:

> interests emerge within the context of a particular belief system and historical experience. This both neo-realists and neo-liberals, in general, fail to take sufficient account of. Indeed, national interests and foreign policy behavior are responses to environmental constraints that are normative and ideational in nature, as well as being structural and material. They are not based simply on a rational calculus of utility maximization.[49]

I concur. Unfortunately, though, Lowi's own substantive conclusions betray this sensitivity to history, ideology and complexity. One cannot simultaneously argue that state action is wholly determined by a national interest in maximising state power, and also that it is informed by 'history, culture and ideology'. Boxed in by her realist framework, Lowi thus ends up depicting co-operation as a function, not of any historical, cultural or ideological factors, but simply of the combined effects of resource need and relative power. The consequence of this is that while Lowi's empirical work is richly historical, her theoretical conclusions are quite the opposite, being essentially ahistorical. This should not surprise us. We have already seen that realists view the workings of international politics as 'depressingly uniform' – one leading realist even questions whether 'twentieth century students of international relations know anything that Thucydides and his fifth-century [BC!] compatriots did not know about the behavior of states'.[50] The result, as Robert Cox writes in critique of Kenneth Waltz (though his words apply equally to Lowi), is that:

> History becomes for neo-realists [and realists generally] a quarry providing materials with which to illustrate variations on always recurrent themes. The mode of thought ceases to be historical even though the materials are derived from history. Moreover, this mode of reasoning dictates that, with respect to essentials, the future will always be like the past.[51]

One problem, then, with Lowi's account is its lack of attention to history and to the ways in which 'national interests' get constituted (by whatever historically emergent cultural, political and economic factors). The second problem can be stated much more briefly: namely, that Lowi's analysis suffers from its exclusive focus on the natural water resources of the Jordan, Euphrates, Nile and Indus rivers. Lowi presents these rivers as finite resources, with fixed natural flow and discharge rates, which in one sense they are of course. What she does not tell us, however, is that many states, in the Middle East as elsewhere, get much of their water supply either by recycling and reusing their apparently 'finite' resources, or by importing water surreptitiously from elsewhere. Lowi, for instance, mentions nothing of wastewater recycling which, as we have seen, is already widely practised in Israel; she says nothing of the fact that the major part of some Middle Eastern states' total water needs are met from rain falling upon Europe and North America; and she says nothing of the options of desalinating water, or importing it in bulk. In ignoring these nascent and potential sources of water, Lowi ends up painting us a quasi-Malthusian picture of international water politics, one that implicitly assumes, even if this is never made explicit, that the 'available water resources are simply limited'.[52] She portrays the Jordan basin as a finite natural theatre for political activity, yet achieves this only through an elision of the multiple sources of water on which Middle Eastern states already rely – through an elision, that is, of the political economic means through which waters are always produced. The latter, it should be said, is typical of and congruent with realism's general insensitivity to the political economic face of international politics.

The consequence of these twin deficiencies is that Lowi ends up giving a mistakenly conservative rendition of Jordan basin water politics. As she says (writing before the onset of the Oslo process) of the West Bank's groundwater resources:

> As long as the *status quo* persists and Israel remains hegemonic both in the occupied territories and in the central Middle East, there is no reason to fear for that critical source. However, the dominant conservative view is that if Israel loses or rescinds political and military control of the West Bank, more than one-quarter of the country's water supply, and hence its national security, could be threatened. Besides, the state cannot be assured that it will remain hegemonic. At some future date, Israel may no longer be in a position to utilize its power resources to avert threats to its security.[53]

This apologist conclusion is the inevitable product of Malthusianism plus realism. By virtue of her portrayal of the available waters as a set of finite resources, Lowi ends up presenting water scarcity as 'a material constraint to survival', and a vital national security issue.[54] Water conflicts are, in Lowi's account, struggles to keep or improve one's share of a strictly limited body of resources, and hence they are – although she never makes this explicit – of an essentially zero-sum nature. In this Lowi would no doubt concur with Frey and Naff when they assert that 'scarcity of water is always a zero-sum security issue and this creates constant potential for conflict': the Middle East's water resources are naturally scarce, they set constraints upon human consumption, and it is for this reason that conflicts over water are so endemic.[55] But water resources, as argued in the previous chapter, are not finite; and the final constraint for states is not the natural availability of their water resources, but their political economic capacity to reuse, recycle, import, desalinate and so on – their ability to marshal water resources across time and space. Equally, Israel's national security would not necessarily be threatened were it to rescind control of some of the Jordan basin or

West Bank's relatively plentiful supplies. To the contrary, seen in the political economic light enunciated here, Israel could quite feasibly grant the Palestinians and neighbouring Arab states a much greater share of the region's water resources. As one leading liberal Israeli water expert, Saul Arlosoroff, observes of the Israeli–Palestinian water conflict: 'The whole issue is about 100 mcm in the foreseeable future, and 100 mcm desalinated from the sea is $100 million, $100 million when Israel's GDP is already $100 billion. That makes it 0.1 per cent of GDP. So from an economic or financial point of view, it's irrelevant, water is irrelevant.'[56] The question inevitably arises, given this, of why the water issue continues to be such a bone of contention between Israel and the Palestinians. As will shortly become apparent, this is not because of some mythical national interest in securing as much water as possible – just as Jordan does not have objective national interests, so the same is true of Israel – but because of certain historically emergent features of the Israeli polity, economy and society.

LAND, LABOUR AND ISRAELI WATER POLICY

Liberal optimism tends to be characterised, as we have already seen, by an insensitivity to politics, and to this extent at least, Lowi's realism provides much the more convincing explanation of Middle Eastern water politics than does the idealism-cum-functionalism of Dolatyar and Gray. In other respects, however, the similarities between Lowi's realism and Dolatyar and Gray's liberal-optimism are much more striking than the differences. Sharply contrasting they might be, but each of these accounts is based on certain ahistorical assumptions about the rationality of state action. For Lowi, states act in accordance with the dictates of power and resource need – and they do this because states make rational assessments of their interests based on objective 'national interests'. Likewise for Dolatyar and Gray, state decision-makers share what, following Alam, they refer to as 'water rationality'.[57] Cognisant of the 'mutual hostage situation' in which they find themselves, states do the rational thing and are thereby driven towards co-operation. In both accounts, states are portrayed as acting rationally, in accordance with their objective interests; moreover in both accounts, the task of explanation is understood in positivist terms, as involving the search for general laws and patterns of state behaviour. Yet contrary to positivist wisdom, state behaviour is not reducible to (or explainable as the product of) objectively given interests; state interests, policies and actions are rather the function of structures, struggles and relations that emerge and change over time, and that must therefore be studied with an eye not to universal truths, but to historical specificities. There are no general trans-historical causes of inter-state conflict and co-operation over water resources – only historically particular reasons why some states co-operate, and why others conflict, over shared water resources.

This has already been illustrated in relation to Jordan, whose water policy has historically reflected not just its poor power and riparian positions, but also the Hashemite tradition of accommodation and collusion with Israel. Turning now to the Israeli state, its definitions of Israel's national water needs and interests (and in turn its policies towards its Palestinian and other Arab riparians) also reflect certain historically specific features of Israeli state and society. To understand these, though, we need to go back to the very roots of the Zionist movement and the Israeli–Palestinian conflict.

There is no better place to start in this than with Gershon Shafir's groundbreaking study of early Zionist settlement, *Land, Labour and the Origins of the Israeli–Palestinian Conflict, 1882–1914*.[58] For Shafir, the Zionist movement was and remains a colonial one, and Israel was established as a colonial society – that is, 'any new society established through the combination, to various degrees, of military control, colonization, and the exploitation of native groups and their territorial dispossession, justified by claims of paramount right or superior culture'.[59] For Shafir, the early days of the Israeli–Palestinian conflict involved a specific though unusual sort of colonial encounter. The early Zionists may not have conceived of themselves as colonisers – and indeed generally did not do so – but in terms of the practical task of settling and establishing a presence on the land, they had to resort to methods comparable to, though in many respects different from, those of metropolitan colonial states. They had to get by; they had to earn a living somehow; and they had to fashion circumstances that would be attractive to the remaining European Jewry (and lest this sounds unimportant, less than 3 per cent of the more than two million Jews who left Eastern Europe between 1882 and 1914 chose to immigrate to Palestine).[60] The early Zionists faced difficulties, however, these being that Palestine was already thoroughly settled, and that the native Palestinian Arab population, being generally poorer, constituted a large reserve pool of cheap labour. Capitalist Jewish settlers like Baron Rothschild, who were buying up land in Palestine around the turn of the century, preferred for obvious economic reasons, to maximise their profits by employing cheap Arab labour on typical colonial plantations. The many propertyless Jews who arrived in Palestine with the second aliya (wave of immigration), from 1903, thus faced intractable problems of getting work. Unwilling to lower their wage demands and thus compete directly with Arab workers, these working-class Zionist immigrants instead started invoking nationalist arguments, campaigning for the exclusion of Arab workers from the Jewish labour market. It was as a product of this class conflict, Shafir argues, that eventually an accommodation was reached between Zionist landowners and workers, without which mass Zionist colonial settlement would not have been possible – this being what Michael Shalev has termed 'a practical alliance between a settlement movement without settlers and a workers' movement without work'.[61] Materially, this accommodation involved at least three main things: the development of co-operative settlements (kibbutzim and moshavim) as the characteristic Zionist economic and social form, these settlements serving to protect Jewish workers and indeed the Zionist project from cheaper Palestinian labour, effectively nationalising the land; the emergence of strong pre-state national institutions such as the Histadrut trade union and the paramilitary Haganah which were established to oversee, protect and indeed institutionalise this specific type of colonial project; and the relative exclusion of Palestinian Arabs from these emerging Zionist institutions. It is here, according to Shafir, where lie the roots both of the Israeli–Palestinian conflict, and of Israel's idiosyncratic state, societal and economic forms.

The strengths of Shafir's account can be readily seen by comparing it with traditional sociological 'functionalist' explanations of Israeli society (these being quite different from the equally functionalist approaches to international relations discussed earlier). From such perspectives – which are most famously associated with Talcott Parsons, and in Israeli sociology with Shmuel Eisenstadt – social forms are understood as being marked by a value consensus, this consensus being the glue that holds the social order together, and that explains its individual characteristics.[62] Israeli society, so interpreted, was fashioned above all upon the socialist, agrarian values of the second aliya – these values having been institutionalised within the kibbutzim, moshavim and the Histadrut, and overseen at a high political level by the Labor Party, which dominated Israeli politics for the best part of 30 years after independence. Where such interpretations go wrong, however, is in assuming the existence of a straightforward value consensus amongst the early settlers; in underestimating the tremendous practical and political economic difficulties that these early settlers faced; and in ignoring, in typically 'internalist' fashion, the foundational impacts of the Israeli–Palestinian conflict upon Israeli society (they operate with what Baruch Kimmerling has termed a 'Jewish bubble' model of Israeli society).[63] Shafir's historical materialist argument manages to avoid each of these failings, implying to the contrary that Labor Zionist ideology more legitimated than founded the specific form of the Zionist project. 'The Israeli–Palestinian conflict,' as Shafir himself observes, 'gave shape precisely to those aspects of their society which Israelis pride themselves in being most typically Israeli'.[64]

The relevance of this in the present context is that it helps shed light on the roots of a number of aspects of Israeli water policy – relating to the place of water within Israeli economy and society; to the institutional capacities of the Israeli state; and to the Israeli state's discriminatory policies against its own Palestinian population. In the first case, it has often been observed that water has long occupied an incredibly important role within the imagination and practices of Israeli society, being of no less than 'ideological' significance.[65] As it developed during the early twentieth century, Zionist thought came to place great weight on agricultural activity, and thereby also on water. In 'exile' – so the nationalist narrative had it – Jews had been landless urban-dwellers, 'alienated' from their 'promised land' as well as from land and labour in general. Zionism thus came to advocate not only a return to Palestine, but also the redemption of the Jewish people through agrarian physical labour, and the transformation and rebirth of the 'wasteland' of Palestine into a 'land of milk and honey'.[66] Control of the region's waters became in turn a 'primary goal' of the early Zionist leadership – it is 'of vital importance not only to secure all water resources already feeding the country, but also to be able to conserve and control them at their sources,' declared Chaim Weizmann at the Paris Peace Conference of 1919 – and the development of these waters became a primary aim of the Yishuv as a whole.[67] Numerous swamp drainage and well-drilling programmes were undertaken, with much of the latter work being conducted by Mekorot, established in 1938, and still to this day responsible for the construction and management of Israel's water systems. Ambitious plans were developed for the transfer of water from the upper Jordan River to the Negev Desert, and also from the Mediterranean to the Dead Sea, and following independence in 1948, the first of these water supply projects started to be put into effect.[68] In 1953, work began on a conduit from above the Sea of Galilee, which was to become the National Water Carrier. A further major pipeline, conveying water from the Yarkon River to the northern Negev, had been completed by 1955, and with the completion of the National Carrier, water was henceforth being conveyed for a distance of over 200 km from the upper Jordan, thereby enabling the expansion of agricultural activity in the south of the country (see Figure 2.1).[69] These investments were not subject to strict economic criteria. To the contrary, at the root of these and other projects lay the central role of agriculture and hence water within Labor Zionist ideology and practice.

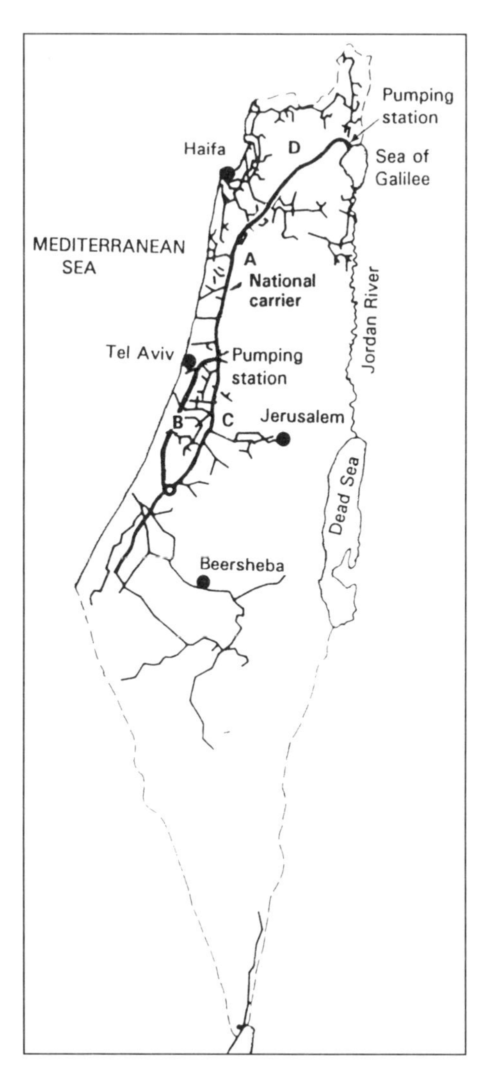

Figure 2.1 Israel's National Water Carrier (reproduced, with permission, from Daniel Hillel, *Rivers of Eden: The Struggle for Water and the Quest for Peace in the Middle East* [1994], courtesy Oxford University Press)

Some have interpreted this as indicative of the formative influence of ideology upon Israeli water policy and the water conflict with the Palestinians. Alwyn Rouyer, for one, has recently argued that the 'origins and perpetuation of the water conflict are best explained not so much by... theories of international politics' – and in this he is surely right – 'but by internal Israeli policies and founding myths which have shaped both past and current Israeli water and agricultural policies... the Israeli–Palestinian water dispute has its roots in internal Israeli politics, political myths and public policy'.[70] Yet if we are to follow Shafir's materialist account of the roots of the Israeli–Palestinian conflict, then it can more accurately be said that Israeli water policy is founded not primarily in ideology, but in the practical territorial imperatives of land and labour within the specifically Israeli–Palestinian context. The 'founding myths' of which Rouyer writes were not in themselves foundational, but were a consequence of the Zionist struggle to colonise the land of Palestine.

Following independence, the same intimate relationship between colonising the land and controlling regional water resources continued to hold. Israel embarked on a policy of establishing Jewish settlements in areas of Arab majority (such as Galilee), and also in peripheral areas of the country (alongside the borders with Syria, Lebanon and Jordan, and in the Negev). Saul Arlosoroff is candid in this regard:

> The whole philosophy of the Zionist movement was that you maintain control of the land, over your country, by working there and being there. There's no doubt that if they move out of the border with Lebanon, somebody else will be there, and that somebody is Arabs, not Jews, and the government of Israel doesn't want Arabs to be there on the border, because then the border will move further and further south. The same is true in the Negev.[71]

For practical economic reasons, this land use took on a predominantly agricultural form. Says Arlosoroff of the areas bordering Lebanon, people living there 'can only grow poultry or horticultural fruit... there is no other alternative; they cannot go from the upper mountains there and work in the cities'.[72] The problem lies in the fact that many of these peripheral areas are very water scarce: thus large supplies of affordable water for irrigation would be (and have remained) 'indispensable' for the 'survival' of such peripheral communities.[73] To maintain a viable presence throughout the land of Israel – a security imperative, no less – water would thus have to be both heavily subsidised, and conveyed to wherever it was required by the state's Jewish populace. Water, as Israel's third Prime Minister Levi Eshkol put it, was 'the blood flowing through the arteries of the nation.'[74] Indicatively, agricultural interests came to assume the dominant role in water policy-making. The Water Commission (responsible for overall regulation of the water sector) was placed under the authority of the Ministry of Agriculture, where it remained until 1996; equally the Water Commissioner was always, until 1990, a political appointee from the agricultural establishment.[75] Both Eshkol and David Ben Gurion were personally involved in water planning, with Eshkol having been the first director general of Mekorot, Israel's national water authority.[76] For these leading lights of Labor Zionism, water was a weapon in a territorial struggle. As during the pre-state period, the practical imperative of controlling and securing a physical presence on the land continued after 1948 to lie at the root of Israeli water policy.

Agriculture is no longer of the economic importance that it once was in Israel, simply by virtue of its developed high-tech manufacturing and service sectors. By 1990, agriculture made up less than 3 per cent of Israel's GDP, and 4.2 per cent of its workforce.[77] Moreover, due to increases in domestic demand, a much smaller proportion of Israel's water supplies are now allocated to agriculture: where in the first three decades of Israel's existence, 80 per cent of its water supplies were used for irrigation, that figure now fluctuates at around 60 per cent (with much of this being recycled domestic water).[78] Israel's basic food staples are, as we saw in the previous chapter, now largely imported from the US. Nonetheless, agriculture is still heavily subsidised, to the extent that it is sometimes claimed to be of negative economic value: 'oranges and grapefruits which are grown in Israel [and] are sold abroad are essentially exported water,' claims one Israeli commentator; 'Israel can import oranges from Europe for less money than it costs to farm them here (if the farmers had to pay the real price for water)'.[79] And this is because while agriculture is economically insignificant, in territorial-political terms it is still deemed of utmost importance. 'The rural and agricultural sector in Israel discharges a national and social responsibility in dispersing population [and] populating frontier regions,' reported the Ministry of Agriculture and Rural Development as recently as 1997.[80] Israel's 'water needs' follow directly from this. Indeed as former Water Commissioner Meir Ben Meir admitted in that same year, 'were it not for the ideological and practical necessity to cultivate and irrigate land, Israel would not have a water problem'.[81]

If the colonial basis of Israeli state formation casts light on the importance of water and agriculture to the Israeli state and society, it also draws attention, more broadly, to the strength of the Israeli state

– which became powerful not simply in relation to neighbouring states, but also in relation to Israeli society, territory and economy. Ever since its inception Israel has been, in Joel Migdal's terms, a 'strong state', overseeing a Jewish society that, at least at first, was not marked by alternative centres of power and resistance that challenged its rule-making and distributive authority.[82] Just as the pre-state institutions of the Yishuv played a central role in structuring the emerging Jewish economy, excluding Palestinians from the protected Jewish labour market and directing investments in line with national and colonial priorities, so the same pattern continued after independence. The Israeli economy became dominated by the state. The state controlled not only land and labour, but also the main flows of capital into the Israeli economy, becoming the beneficiary first of Holocaust reparations and later of Cold War-related military and economic aid from the US. Indeed in its first 20 years, around three-quarters of all capital imports into Israel were received by the state, with the state in turn financing nearly two-thirds of total capital formation.[83] Given this, the public sector became an extremely influential distributor of rents, these being allocated in accordance with developmental imperatives, and primarily oriented towards the expansion of agriculture and the provision of (largely agricultural) employment for new Jewish immigrants.[84] The state also, of course, became militarily and coercively powerful – both internationally, where it quickly established Israel as a regional hegemon, but also internally, where the military remains so influential that the distinction between 'military' and 'civilian' sectors barely applies. Israel, as Uri Ben-Eliezer has cogently argued, is a militaristic society, where military and national security discourse occupy the formative role in public life, and in which the views of the military are hegemonic to such an extent that – by contrast with praetorian states – military coups are simply not necessary.[85] Both institutionally and in terms of public discourse, the Israeli military has thus had a formative and preponderant influence upon Israeli society. In each of these respects, the Israeli state has long been the dominant institutional feature of Israeli society.

The significance of this for us is that the Israeli state has been well able to oversee the development of its water resources in accordance with national priorities. Policies without power are of little use; but the Israeli state had both a clear policy of expanding agriculture and dispersing the Jewish population, and the domestic legitimacy and institutional and economic capacity to put those policies into effect. Under the draconian terms of the 1959 Water Law, all water resources became defined as 'public property', 'subject to control by the state'.[86] Moreover, the management of water resources and systems became the preserve of a centralised ensemble of state and para-state institutions.[87] At the apex of this ensemble was (and remains) the Water Commission and its Commissioner, responsible for allocating production and use licences, imposing sanctions against non-compliance, administering water rates and subsidies and overseeing hydrological planning – and thus being responsible for the vast majority of water-related administration, and having an influential role in directing, interpreting and implementing policy. Below it are a number of further institutions: Mekorot, the National Water Authority, which is responsible for the construction, maintenance and operation of the Israeli water network, which supplies the larger part of the water consumed within Israel; Tahal, which was defined by the 1959 Water Law as responsible for planning and designing water facilities, and developing supply and demand projections; and the Israeli Hydrological Service (IHS), which is responsible for the monitoring and modelling of surface and subsurface water resources. Municipal authorities also play a key role in the local distribution of water, typically receiving it from Mekorot and thereafter supplying it to consumers. There have of late been a number of changes in this institutional regime, as well as broader changes in the nature of Israeli state-economy relations (these will be discussed in subsequent chapters). Nonetheless, what were established shortly after independence were a centralised set of public water institutions that were financially well-supported and clearly oriented towards the fulfilment of national objectives. During the construction of the National Water Carrier and its branches, that is between 1950 and 1970, state investment in water infrastructures represented between 3 and 5 per cent of gross capital formation, an astoundingly high level.[88] While this level of investment no doubt reflected the importance of water and agriculture within the Zionist colonial project, it also testified to the administrative and developmental strength of the young Israeli state.

Finally, and more broadly still, the colonial nature of Israeli state formation also serves to highlight the essential character of the conflict with the Palestinians. Fundamentally this was a conflict over land, from which the Palestinians were increasingly excluded – at first through institutional and economic means, but later through coercion and expulsion. Prior to 1948, Zionist strategy was largely premised on the Jewish National Fund buying up Arab land, often from absentee landowners, and the nationalist institutions of the Yishuv excluding Palestinians from working on these now Jewish lands.[89] Increasingly, though, these exclusivist institutions were backed up in their work by the Haganah and other Zionist paramilitary organisations, as Mandate Palestine became the site of a three-way conflict between the British colonial authorities, Zionist settlers, and the often landless Palestinians. The Yishuv, and Zionist strategy, became increasingly militarised, to the extent that during the lead-up to the war of 1948–49, plans were laid for the mass expulsion and transfer of Palestinians, clearing the way for the formation of as pure a Jewish state as possible.[90] In the event, over 700,000 Palestinians fled or were expelled. Only 150,000 Palestinians remained within the new state of Israel, the Palestinian population having been reduced from a 2:1 majority in Mandate Palestine, to a small minority of around 12.5 per cent of the population of Israel.[91] Israel now comprised 73 per cent of the total area of Mandate Palestine.

These new demographic and territorial realities made it relatively easy for Israel to incorporate the remaining Palestinian minority into its economy and society. They became entitled to Israeli citizenship, and were partially integrated into the Israeli economy as a cheap labour force who no longer posed such a threat to the Zionist colonial project.[92] Nonetheless, the struggle for the land continued. Land belonging to refugees, or designated for military or Jewish settlement purposes was confiscated; land was bought from Palestinian landowners by the Israel Land Administration; and Palestinian physical development was retarded. Palestinian agriculture, meanwhile, was 'all but destroyed by the state's land and water policies'.[93] Water was (and continues to be) allocated disproportionately to the Jewish sector, such that in 1988, for instance, Palestinians worked 19 per cent of the land in Israel while receiving only 2.7 per cent of the water.[94] Domestic water and sanitation services for Palestinians also remained far below those for the country's majority Jewish population, with most Palestinian towns and villages still being without central sewage networks, for instance.[95] Until 1966, areas of Arab concentration were subject to a military administration that continued to restrict in practice those few freedoms that they attained through Israeli citizenship. In its relations with its Palestinian minority, then, the Israeli state practised an institutionally oiled form of apartheid, one that both followed from and served to consolidate the colonial nature of the Israeli state.[96] As we shall see in the next chapter, Israel's post-1967 policies in the West Bank and Gaza have continued in much the same vein.

The above represents a necessarily cursory overview of the historical roots of Israeli water policy. My assumption here has been that in order to explain and understand patterns of international conflict and co-operation over water resources, we need to look at the various ways in which the specificity of a country's politics and political economy leads to the uptake and institutionalisation of certain definitions of state needs and interests *vis-à-vis* water. Against realism, my assumption has been that patterns of international conflict and co-operation are not reducible to objectively given state needs or national interests. More empirically, I have argued that Israeli water policy has historically reflected the specific character of the Zionist colonial encounter with the Palestinians – that the importance of agricultural and water development within Israel, the institutional capacities of Israeli water institutions, and the Israeli state's discriminatory land and water policies towards its Palestinian minority all have their roots in the early period of state formation. Water is generally of only minimal political economic importance to Middle Eastern elites. In the Israeli case, however, the specific pattern of colonial settlement fostered the emergence of a Labor Zionist elite and a state-led public discourse which viewed water as of inordinate importance, and that primarily viewed water issues through a nation-building and national security lens. The consequence of this has been that the Israeli state has historically been very activist in exerting and extending control over regional water resources – by constructing what is now the most integrated national supply network in the world; by resorting to force to prevent the Arab League's diversion of the Jordan River; by invading and occupying the Golan Heights, possibly in part for hydro-strategic reasons; by being unwilling to concede control of the shore of Lake Tiberias to Syria, out of fear that even a full peace treaty would not stand in the way of Syrian exploitation of this vital national reservoir; and by minimising its own Palestinian population's use of national water resources.[97] For much the same reasons, the Israeli state has also strongly discriminated against the Palestinians in the Occupied Territories, minimising their water use so as to make as much water as possible available for Jewish Israel and the Zionist project. Having explored some of their long-historical contexts, it is to these more recent matters that we now turn.

CHAPTER 3

The West Bank Under Occupation: Technologies, Policies and Power

In June 1967, Israel invaded and occupied the remaining 27 per cent of Mandate Palestine (the West Bank and Gaza Strip), along with the Syrian Golan Heights and Egypt's Sinai peninsular (Figure 3.1). The startling success of Israel's military conquests was to prove a double-edged sword. For while in 1948–49, the long drawn-out conflict had seen the flight of over 700,000 Palestinians, in 1967 only 355,000 Palestinians and 80,000 Syrians fled their homes, and Israel suddenly found itself administering an additional Arab population of 1.5 million people. For Israel this situation posed new demographic, political and political economic challenges – ones to which the Israeli state and Israeli society responded by extending and deepening those exclusivist apartheid practices first developed against the country's own Palestinian population.

This chapter both overviews these post-1967 developments, and details the various means by which the Israeli state extended control over the West Bank's water resources in accordance with the nation-building and territorial imperatives already discussed. I begin by describing the broad contours of Israel's occupation policies, depicting these as an extension, and indeed culmination, of the Zionist movement's colonial encounter with the Palestinians. I then consider Israel's water policies in the Territories, and the legal, institutional and also technological means through which these policies were exercised, and through which they became possible. It was suggested in conclusion to chapter 1 that the tasks of structural explanation and critique demand analyses that, rather than assuming the separate existence of technical and political problems, instead view water crises as political economic

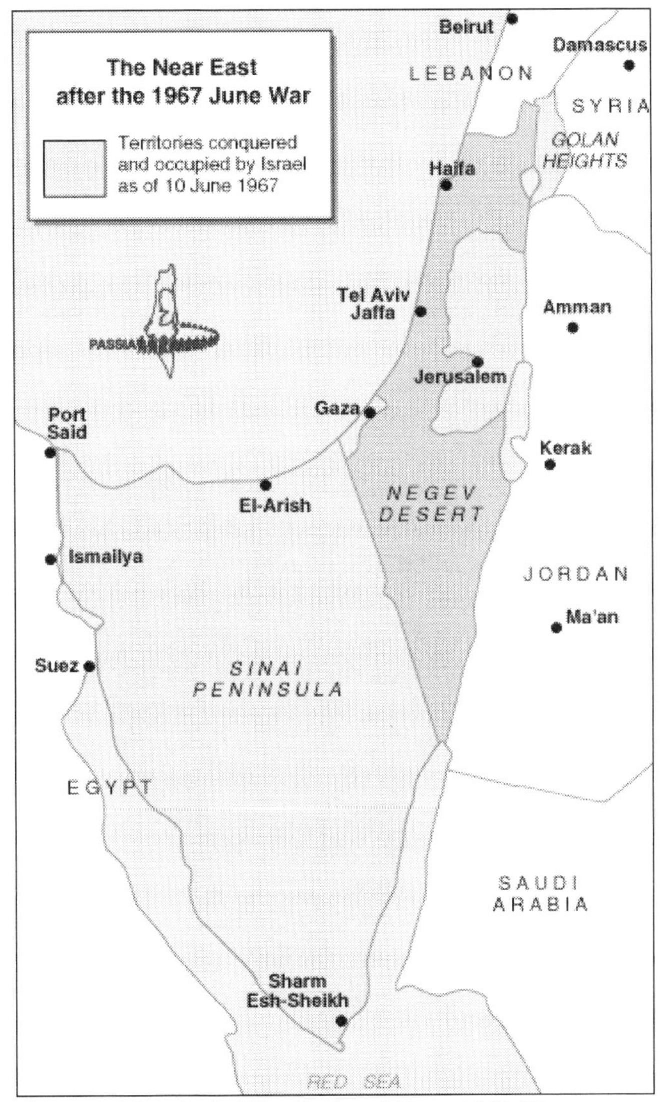

Figure 3.1 Israel, the West Bank and Gaza, 1967–93 (reproduced, with permission, from the Palestinian Academic Society for the Study of International Affairs [PASSIA], www.passia.org)

issues which have both macro-level and micro-scale techno-political causes. In line with this, the previous chapter developed a macro-scale political economic analysis of the historical roots of Israeli water policy. This present chapter, by contrast, develops micro- and macro-scale analyses alongside one another, considering both the political economy of colonisation in the Territories, as well as the micro-physical means through which Israeli policy was actualised. Technologies in this case are not simply noticeable by their absence, I argue, but are part and parcel of the workings of Israeli power.

COLONISATION CONTINUED

Between 1967 and 1993, there were five central planks to Israeli policy in the Occupied Territories: the extension of territorial colonisation through settlement-building, land confiscation and bypass road construction; the economic absorption of the West Bank and Gazan economies into the much larger Israeli one; the creation of a dual institutional and legal system that applied different laws to Palestinians and Israeli settlers, and that directed investment much more towards the latter than the former; the search for compliant local leaders and patrons who would do Israel's bidding; and, when this failed, the increasingly harsh repression of the occupied Palestinian population. We briefly consider these in turn.

In the years immediately following the 1967 war, the direction and indeed depth of Israel's colonial ambitions in the Occupied Territories remained uncertain. While Levi Eshkol had declared just a week after the war that Israel would not return to its former post-1949 borders, this did not immediately presage any renewed state-directed colonisation drive.[1] Within the Israeli cabinet, there was significant disagreement between 'hawks' such as Defence Minister Moshe Dayan, who favoured the establishment of Jewish settlements in the West Bank, and 'doves' such as Pinhas Sapir, who advocated the return of most of the Territories.[2] Meanwhile within the country at large, the Labor Zionist movement had lost some of its pioneering colonial ambitions.[3] It was not until the emergence of the religious-nationalist Gush Enumin movement in the mid-1970s, and more particularly the re-election of Menachem Begin and his Likud coalition in 1981, that Israel's colonisation of the Territories seriously got under way.

During the ten years following the 1967 war, Israeli settlement-building in the Occupied Territories was officially informed only by security considerations, though increasingly also in practice by religious-nationalist ideology. The Labor government's Allon Plan of 1969 called for the development of a chain of settlements in the Jordan Valley such that this strategically important area could not be returned to Jordan, and the establishment of settlements along the Green Line to the north and south of Jerusalem, thus enabling the expansion of Israel's heartland; it also stipulated that settlements should not be constructed in the heavily populated Palestinian areas of the central West Bank.[4] In accordance with this, most of the settlements established before 1977 were located in the Jordan Valley – these being Labor Zionist settlements, set up as agricultural collectives in a direct continuation of pre-1967 policies. In addition to these, though, an increasing number of settlements were being established outside the bounds of government policy by the Gush Enumin (Bloc of the Faithful) movement. Inspired by a belief in the Jewish people's national-religious rights to the whole of Eretz Israel, Gush Enumin established settlements in Palestinian populated areas, with the most provocative of these being in and around the city of Hebron.[5] These ideological settlements were markedly different from those established under the Allon Plan, sharing none of the agrarian, socialist and Labor Zionist emphases of the latter. In other respects, though, Gush Enumin's settlement-building represented a continuation of longstanding Zionist practice. As Rabbi Levinger, who led the settlement of Palestinian Hebron, once claimed, Gush Enumin was no less than 'the direct and legitimate offspring of the pioneers of Zionism'.[6] The actors and ideology behind Israel's territorial drives had changed, but the struggle to colonise the land was nonetheless an inheritance from the pre-state days.

Yet important as this development was, the main settlement drive in the Occupied Territories began only in 1981, with the second term of the Begin government. For the first time then, the government adopted a policy of redirecting some of Israel's growing urban sprawl to the West Bank, and establishing there a number of large-scale dormitory settlements.[7] Under the terms of a Jewish Agency plan formulated by Mattiyahu Drobless, the Jewish population of the West Bank would be brought up to 1.3 million by 2010.[8] To support this process, massive incentives were provided for settlers, the majority of whom moved to the West Bank for economic rather than (Labor Zionist or national-religious) ideological reasons. House prices and services were subsidised, extra social benefits were made available to those relocating to West Bank settlements, and numerous bypass roads were constructed, enabling settlers to commute to Jerusalem and beyond without having to pass through Palestinian population centres.[9] Settlement continued in this fashion throughout the 1980s, and even continued following Labor's return to power in 1992, despite Yitzhak Rabin having campaigned on a promise of 'changing national priorities' and ending all other than 'natural growth' of the settlements. By 1993, the West Bank (including East Jerusalem) had become home to 280,000 settlers.[10]

A second and less widely recognised dimension of Israel's occupation policy lay in the dependent incorporation of the West Bank and Gaza economies into the much larger Israeli one.[11] Palestinian agricultural activity in the Territories was actively restricted, especially where it competed with Israeli agriculture – partly through the introduction of quotas to limit and in some cases prevent the sale of produce to within Israel, and partly (as we shall see) through control and discriminatory pricing of water supplies. Manufacturing development was also intentionally stifled. As a result the Palestinian economy came to depend more and more on poorly paid and non-unionised day labour within Israel, such that during the late 1980s, as many as 165,000 Palestinians would journey each morning from the West Bank and Gaza (and many thousands also worked in the construction and maintenance of Jewish settlements, and in Jewish industrial estates in the West Bank).[12] In addition to constituting a captive labour force, the Palestinians of the West Bank and Gaza also became a captive market for Israeli goods, with the Territories becoming, by the early 1990s, Israel's second biggest market after the US.[13] Until the early 1980s, individual prosperity in the Territories did increase, in part because of the relatively high wage rates within Israel, but also because many skilled Palestinian workers had migrated to the Gulf states, and would regularly send remittances back to their families in the Territories. However, with the collapse of the oil-driven boom in the Gulf during the early 1980s, these remittances dried up, with inevitable knock-on effects for the economies of the West Bank and Gaza. While personal prosperity at least for a while increased, this was within the context of growing absorption into and dependency upon the dominant Israeli economy.

Third, Israel's colonial policies in the Territories were supported through the creation of a dual legal and institutional regime.[14] Just as Israel's own Palestinian population were, until 1966, denied formal citizenship rights by being placed under military administration, so too the Palestinians of the West Bank and Gaza were made subject to military government. They were wholly denied Israeli citizenship rights, enabling Israel to avoid coming to terms with the contradiction of wanting to be both a Jewish and a procedurally democratic state, while having over two million Palestinians – now four million Palestinians – under its control. Palestinians were made selectively subject to some Israeli laws and regulations, such as Israeli taxation levels. Nonetheless, all 'powers of government, legislature, appointment and administration in relation to the region or its inhabitants' were placed in the hands of a Military Government and Governor (and later in the hands of the renamed Civil Administration), and the Palestinians of the West Bank and Gaza became largely subject to an apparatus of Military Orders.[15] These same Military Orders did not apply, however, to Israeli settlers residing in the West Bank and Gaza, who instead exercised the full range of political, social and economic rights available to all other Jewish Israeli citizens (indeed in many respects they had more economic and social rights than other Israelis, as we have already seen); and neither did they apply to non-settler Israelis present within the West Bank or Gaza. Israeli settlers in the Occupied Territories would be tried by civil courts, while Palestinians would instead be tried by military ones. Moreover, Israeli settlers received public funds from the Ministry of Interior, while Palestinians were dependent upon the Military Government, to obvious effect: where in 1983/4, for example, per capita expenditure in public services in the Israeli settlement of Kiryat Arba was $260, in neighbouring Hebron it was only $54; where in the regional council of Samaria it was $568, in Jenin sub-district it was only $12; and where in Mateh Benjamin it was $406, in nearby Ramallah sub-district it was only $8.5 per capita.[16] This despite the fact, as already noted, that Palestinians were subject to the same tax regime as operated in Israel. Overall, the occupation was – at least until the onset of the intifada in 1987 – an extremely profitable venture for Israel. Israel had developed not just an apartheid system of 'legal dualism' in the Occupied Territories, but also a system for discriminating in the distribution of services and investments.

In addition to this, Israel also endeavoured to develop systems of limited self-government under which Palestinian clients would manage local civilian affairs within the broader framework of military occupation and colonisation.[17] While political organisation within the Territories was totally prohibited, municipal authorities and other service providers inherited from the pre-1967 period were generally left well alone. In the West Bank, the local authorities were at first mostly led by pro-Jordanian elites, and continued to receive Jordanian government subsidies. Only during the late 1970s did the municipalities become increasingly led by Palestinian nationalist PLO supporters, this leading to intensifying confrontation with Israel, and in turn both to Israel's search for new clients in what they hoped would be a more amenable rural population (the Village Leagues scheme), and to the Israeli invasion of Lebanon (the objective of which, according to the then Defence Minister Ariel Sharon, was to 'solve the problems of the West Bank and Gaza' – by crushing the PLO militarily at a time when it was losing control politically).[18] Yet irrespective of these developments, local councils and municipalities continued to play a key role throughout the occupation period, enabling the Israeli authorities to minimise their everyday functional and administrative contacts with the Palestinians.

Finally, as the Palestinian population of the Territories became more nationalist – this resulting not only from external PLO agitation, but also from Israel's creeping colonisation of the West Bank, and during the 1980s from the drying up of Gulf boom remittances – Israeli occupation policy became increasingly repressive. The details need not occupy us too long here: suffice to note, by way of illustration, the reported advice of Ariel Sharon and his Chief of Staff Rafael Eitan, that soldiers should 'cut off the testicles' of demonstrators, and that 'the only good Arab is a dead Arab'; the practice of breaking the wrists of those Palestinians who observed daylight saving time during the intifada (this practice, called for by the Unified National Leadership of the Intifada, being such that the Occupied Territories and Israel ran on different times for certain periods of the year); and the broader pattern of deaths, injuries, detentions, deportations and torture that were part and parcel of the Israeli response to the Palestinian call for an end to occupation.[19] Notwithstanding all this, the intifada of course continued, and became increasingly militarised under the growing influence of Islamist movements. It was within this context, as we shall see in the next chapter, that Yitzhak Rabin and Shimon Peres sought to co-opt Yasser Arafat and the Tunis-based PLO for the job of administering and repressing the restive Palestinian population.

PIPELINES AND POLITICS

Each of these Israeli policies in the Occupied Territories find their direct counterpart in Israeli water policies in the West Bank. Overall these policies followed the long-established discriminatory principle of restricting Palestinian consumption so as to maximise the amount of water available for Jewish Israeli purposes. More specifically, they aimed to prevent Palestinian development of the trans-boundary Western Aquifer, which was already being exploited from inside the Green Line, and was (and still is) one of Israel's key sources of freshwater; to curtail use of water within Palestinian agriculture (since as the Director General of Tahal observed, 'every dunam cultivated or irrigated by Arabs will lead to the thirst of a dunam in Israel'); to restrict Palestinian domestic water consumption; and to aid the colonisation of the Territories by providing settlements with regular and plentiful water supplies.[20] These aims were fulfilled through a range of institutional, legal and techno-political means.

As we have already seen, even before the end of the June 1967 war, all 'powers of government, legislature, appointment and administration' in the West Bank and Gaza had been placed in the hands of a Military Governor. Under him, a Water Officer was entrusted with full control over water-related matters within the West Bank.[21] This Israeli Water Officer and the Water Department of the Military Government (later the Civil Administration) became responsible for the allocation of permits and licences, and effected policy dictated by the Israeli Water Commission and also the Ministry of Defence. Under Military Order 158 of October 1967, no person in the West Bank was 'allowed to establish or own or administer a water installation...without a new official permit,' while under Military Order 291 of December 1968, all water resources were declared to be public property, as in Israel itself – all privately and municipally owned Palestinian water facilities henceforth came under the legal gaze of the Israeli state.[22] Military Orders also granted Israeli water officials the power to refuse permits, and to revoke and amend licences, 'without giving reasons'.[23] Meters were installed on all existing wells, and quotas were rigorously enforced, with excess abstraction punishable with heavy fines.[24] Unsurprisingly, permits for new wells and for the repair of existing ones were routinely denied. No permits were issued to Palestinians for the drilling or repair of wells into the Western Aquifer.[25] And agriculture was especially singled out: not one permit was granted for agricultural wells during the 1967–95 period and, incredibly, irrigation was not permitted after four in the afternoon.[26]

Quite apart from the Military Government, Israel's parastatal water company, Mekorot, also came to wield significant influence within the West Bank water sector, especially after 1982 when, in line with the newly expansionist policies of Begin's second government, the then Defence Minister Ariel Sharon oversaw the transfer of ownership of all water supply systems in the territories to the company – which paid for these assets (estimated at a value of $5 million) a symbolic price of just one shekel.[27] From then onwards, Mekorot controlled abstraction rates from the West Bank's deep wells, doing so from its central office in Ramla, near Tel Aviv. Under occupation, and especially from 1982 onwards, the West Bank's water resources and systems came firmly under Israeli control.

Palestinians did nonetheless play a key role in the lower-level management of the West Bank water sector. Prior to 1967, this sector had been administered by the West Bank Water Department, answerable to the Jordanian Natural Resources Authority (NRA).[28] In 1967, however, the Water Department was relocated from Jerusalem to a site adjacent to the Israeli military headquarters at Beit El, where it was placed under the authority of the Military Government.[29] Although some Israelis came to work there, the Water Department continued to be staffed mainly by those Palestinians who had previously worked under the NRA. Prior to 1967, the Water Department had fulfilled a wide range of administrative tasks, having had its own drilling rigs, for instance; after 1967, these fell into disrepair and the Water Department was effectively de-institutionalised.[30] Thereafter the Water Department was only responsible for mundane functional tasks, such as maintaining the West Bank's water network, controlling the volume and flow of water supplied to Palestinian communities (by opening and closing supply valves), and billing (the Water Department would in turn be billed by Mekorot, the water provider).[31] Thus although after 1982 it was Mekorot which owned the West Bank's water supply infrastructure, and controlled abstraction rates from West Bank wells, it was the Palestinian staff of the Water Department who were directly responsible for liaising with Palestinians.

The full significance of this arises from the fact that the Water Department's relations with Palestinians were quite different from those that it had with Israeli settlers. The Water Department was not allowed to close supply valves feeding Israeli settlements, and hence only rationalised supplies to Palestinian communities (this in part explains why Israeli settlements received constant water supplies, while Palestinian communities would be subjected to lengthy cuts).[32] Moreover, settlers were billed by Mekorot rather than by the Water Department, and as in Israel itself, paid for their water at highly subsidised rates, such that the settlers paid much lower rates than their Palestinian counterparts (since Mekorot charged the Water Department at non-subsidised rates. By one source, during the mid-1990s settlers were paying $0.40 per cubic metre for domestic water and $0.16 for agricultural uses, while Palestinians were paying a standard rate of $1.20 for piped supplies for both domestic and agricultural purposes).[33] The Water Department hence functioned as a key institutional interface between the military authorities and the occupied Palestinian population, ensuring that Israel's discriminatory water distribution and billing policy could be effected without any direct contact between Israeli water officials and Palestinian users. It functioned essentially as a client institution. (I should emphasise that my intention in saying this is not to blame those working in the Water Department for their role in effecting Israeli policies. Those living under occupation of course have to earn a living, as is all too evident from the number of Palestinians who work in Israeli settlements. My aim is to draw attention to a structure of relations, rather than to assign any responsibility.)

Palestinian municipalities and village councils also played a key role in local water management.[34] Municipalities and village councils were responsible for maintaining internal networks and for billing individual households within them (forwarding these payments to the Water Authority, which in turn made payments to Mekorot). In addition to this, the larger municipalities such as those of Hebron and Bethlehem operated rotation systems, opening and closing supply valves to ensure that, despite supply shortfalls, water would be received at sufficiently high pressures within at least some parts of their internal networks. In each of these various regards, both the Water Department and the countless municipalities and village councils were absolutely pivotal, not just in managing the West Bank's water sector, but also in effecting Israeli water and occupation policy.

Perhaps the most important means through which Israeli water policies were effected, however, was through the water supply network that was constructed by Mekorot across the West Bank from the early 1980s onwards. Some facilities were constructed prior to this, with the first of the deep wells in the southern West Bank being drilled in 1971, for instance.[35] However, most of the occupation supply network was constructed in support of the new settlement expansion policy set out in the 1981 Drobless Plan, that was being pushed forward by Ariel Sharon and his Civilian Administrator Menachem Milson. The network served primarily to ensure that Israeli settlers received adequate enough water supplies to make them want to live in an occupied territory; but it also functioned to discriminate between Palestinians and Israeli settlers, to integrate the West Bank into Israel's national water supply network in accordance with the Likud's annexationist ambitions, and as an inevitable corollary, to create large supply variations between and within Palestinian communities. The effects of this network are very little commented upon in the existing literature. Al Rouyer, for instance, states that 'the core elements of Israel's water policy in the occupied Palestinian territories were (1) the prohibition of the drilling of new wells or the deepening or repair of existing wells without a permit; and (2) the metering of all wells in order to enforce strict quotas on Palestinian water utilization'.[36] This claim is premised, however, on an overly negative and juridical understanding of power, one that sees

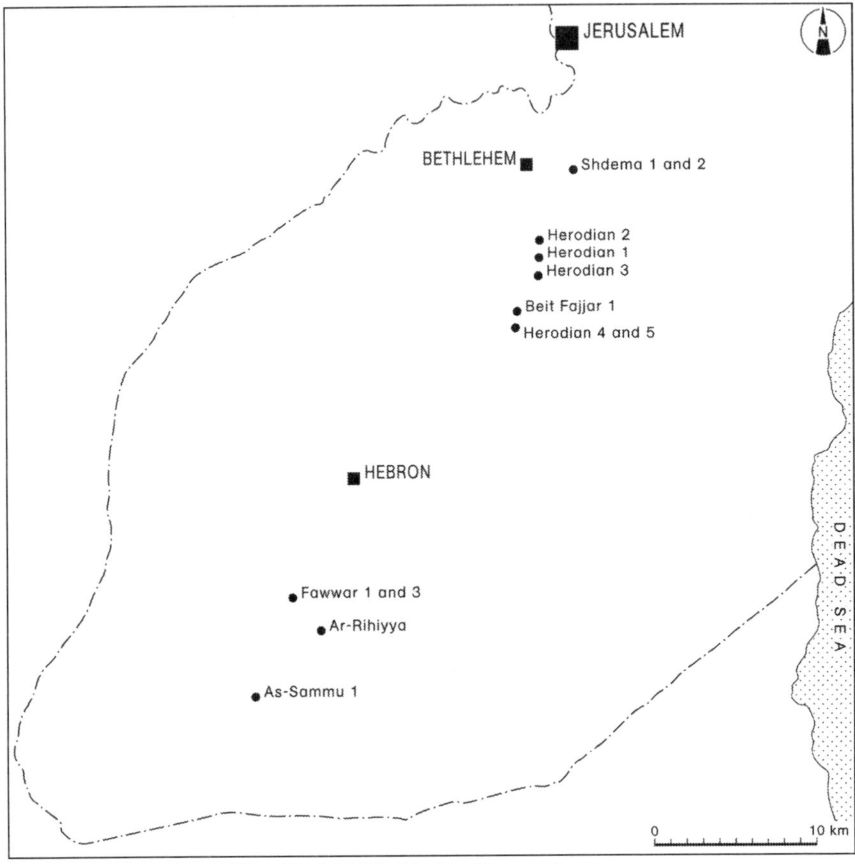

Figure 3.2 Southern West Bank Wells Under Occupation

outcomes wholly as a function of repressive policy, and fails to recognise the productive work of technological systems in effecting state policy and giving form to state power. In truth, the water supply network that had been constructed across the West Bank by 1995 (which was when the occupation formally ended in the West Bank) was the key material-technological cause of the Palestinian water crisis. That this was the case can be clearly shown by tracing some of the flows of water that would have taken place (and that in many cases still do) through the southern West Bank's water systems.[37]

In 1995, the Palestinian communities of the southern West Bank were in receipt of water from 12 deep wells, each of them producing water from the Eastern Aquifer. Six of these were located in the Herodian hills to the south of Bethlehem; another two lay just to the east of Bethlehem, and the remaining four to the south of the city of Hebron (see Figure 3.2). These wells together provided a supply of 13.5 mcm

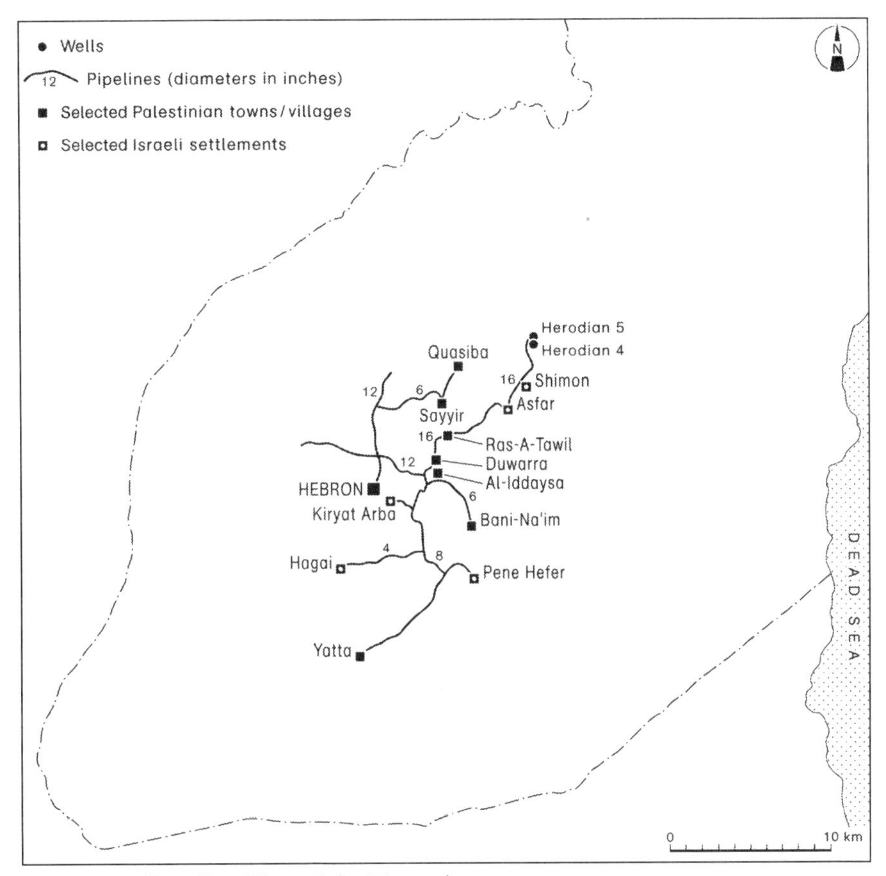

Figure 3.3 Herodian-Kiryat Arba Network

during that year.[38] If we follow, to start with, the waters produced by Herodian Four and Five – which were the southernmost of the six Herodian wells – water from them was pumped southward through a 16-inch diameter and 20 km-long pipeline towards the Israeli settlement of Kiryat Arba (Figure 3.3).[39] Along the course of this pipeline lay several Palestinian villages (Al-Iddaysa, Duwarra, Ras-A-Tawil) and small Israeli settlements (Asfar and Shimon), all of which were connected to it by on-off valves, and by short and small-diameter distribution lines.[40] Just before reaching Kiryat Arba, a 12-inch branch line directed water westwards from this line, through a pump station, to another main line, feeding Hebron from the north; and shortly after this, a further line branched off to the Palestinian town of Bani Na'im, conveying water along a 6-inch pipe to a small concrete reservoir with a storage capacity of 150 cubic metres. However, the larger part of the waters pumped from Herodian wells Four and Five would have continued along the

16-inch pipeline to Kiryat Arba, where they would have passed into, and been stored in, two 1000-cubic metre reservoirs. Much of this water would have been consumed within Kiryat Arba itself. Some of it would, however, have been pumped along an 8-inch line branching to feed the Israeli settlement of Pene Hefer and the Palestinian town of Yatta; and some would have been directed along a 4-inch branch line to the settlement of Hagai.[41]

The pertinence of this fairly typical example is twofold. Firstly, it points to the fact that, as of 1995, Israeli settlements and Palestinian towns and villages were being supplied with water through one and the same network. Israel had constructed a single integrated supply system. The same pattern could be evinced throughout the southern West Bank, with every one of the region's 12 wells supplying both Israeli and Palestinian users. Network integration was not simply internal to the West Bank, however. The two Shdema wells to the east of Bethlehem were part of a network which supplied Palestinian Beit Sahur, the Israeli settlement of Ma'aleh Adumin, and East and West Jerusalem.[42] Many of the Israeli settlements to the south of Yatta were supplied through lines from Be'er Sheva, as were some of the Palestinian villages.[43] Moreover, the entire western flank of the southern West Bank was being fed with water from within Israel, primarily through two 16- and 10-inch pipelines to reservoirs at Allon Shevut, between Bethlehem and Hebron; these were connected to lines supplying Palestinian and Israeli communities as far south as Dhahriyya, and including Hebron (Figure 3.4).[44] Much of the southern West Bank was thus effectively integrated into Israel's national water supply network.

That this integrated supply system also discriminated between Israeli and Palestinian users is less obvious until one considers the relative size of the West Bank's various communities. To return to our earlier example, it is no coincidence that the 16-inch diameter pipeline from the two Herodian wells terminated at and fed an Israeli settlement, rather than a similarly sized Palestinian town or village. As of 1995, such wide-diameter pipelines – and being wider they enabled much more water to be pumped through them – were only to be found supplying Israeli settlements: Kiryat Arba, but also Ma'aleh Adumin to the east of Jerusalem, and Betar 'Illit, Geva'ot and Efrat in the Gush Etzion bloc to the south-west of Bethlehem.[45] What renders this significant is that these Israeli settlements were not nearly the most populous of the southern West Bank's communities.[46] Kiryat Arba – fed by a 16-inch pipeline, but also receiving supplies from elsewhere (on

Figure 3.4 Western West Bank Network

which more shortly) – had a population in 1995 of 5500. Compare this with Bani Na'im, which had a population of 9000, yet received water only through a 6-inch line; with Hebron city, which had a population of 95,000, yet was fed merely through two lines, one 12-inch and the other of 10-inch diameter; and with Dhahriyya which, despite having a population of 15,000, was supplied only through two 6- and 4-inch lines, and received a year-long supply of just 150,000 cubic metres of water, 10 cubic metres per person per year.[47] One finds a similar pattern if one compares population size and storage reservoir capacity. Where Kiryat Arba had two reservoirs, each of 1000 cubic metre capacity (and, besides these, also received water from reservoirs shared with Hebron municipality), the much more populous but Palestinian town of Bani Na'im had just one reservoir of 150 cubic metres.[48] Likewise Dhahriyya had but a single reservoir of 200 cubic metres: a tenth of the storage capacity of Kiryat Arba for a town with three times as many inhabitants.[49] Of the water-stretched Palestinian communities lying close to the Green Line, not a single one of them had any storage capacity (and this includes towns like Ithna with an 11,000-strong population, and Surif with 8000 inhabitants).[50] Indeed within the southern West Bank as a whole, the only reservoirs with greater storage capacity than those in Kiryat Arba were located in Beit Jala (but this reservoir was out of operation in 1995); and just a few miles away, on the northern outskirts of Hebron city.[51] This latter case is worth describing in more detail.

Khaled Batrakh reservoir on the northern outskirts of Hebron was fed, as of 1995, with water from a 12-inch pipeline from the north (which carried water from the Herodian Three and Beit Fajjar wells, and from Israel's national supply network), and it also received water from a further 12-inch line from near Kiryat Arba (this carried water pumped from Herodian Four and Five, as detailed above).[52] The reservoir had a storage capacity of 3875 cubic metres, being the largest in a series of reservoirs which, in principal, were supposed to regulate Hebron's water distribution system. Five metres deep, the reservoir had two outflow pipes (Figure 3.5). One of these, exiting at a height of 2 m from the base of the reservoir, fed into further storage reservoirs, with most of the water contained therein feeding Palestinian Hebron, and a lesser amount being pumped to Kiryat Arba and other Israeli settlements. The second outflow pipe exited from the base of the reservoir, and fed Kiryat Arba alone. Hence whenever the reservoir's water level fell below 2 m – as it often did for long periods each summer – all of the remaining supplies would flow to Kiryat Arba; and Palestinian Hebron, with a

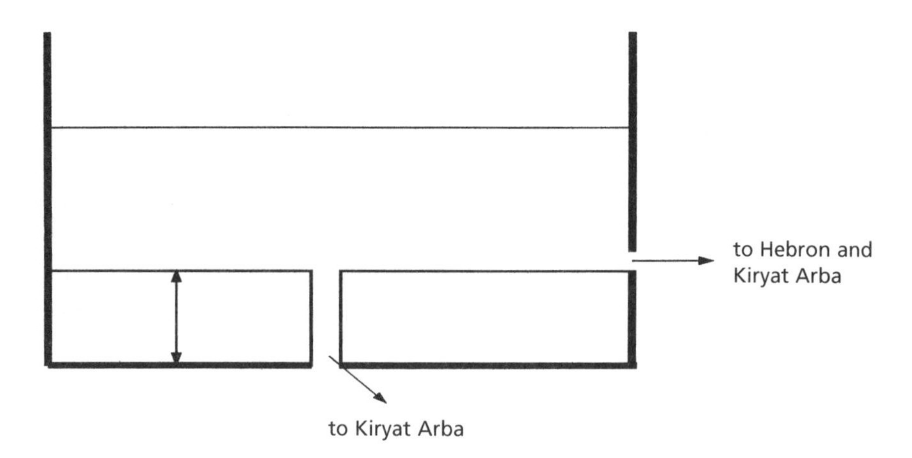

Figure 3.5 Khaled Batrakh Reservoir

population of 95,000 people, would be forced to make do with water from its Fawwar wells to the south, supplied through a single 10-inch pipeline. The result, as briefly mentioned in the introduction, was that Hebron faced long summer shortages, with most areas of the town receiving water for only one day in 20.

This reservoir represents a microscopic instance of the much larger techno-political apparatus which, at the time of the Oslo II Agreement, governed the distribution of water across the southern West Bank. Though the reservoir's fine structure and its specific distributive effects are unique, it is nevertheless paradigmatic of the West Bank's water infrastructure as a whole: it wove together Israeli and Palestinian communities within a single web, conjoining their respective living spaces, while at the same time discriminating sharply between the two, ensuring full and well-regulated supplies for Kiryat Arba's settlers at the expense of Hebron's Palestinians. Khalad Batrakh reservoir was the very epitome of the Israeli state's infrastructure of occupation.

Besides having these powerful integrative and distributive effects, this techno-political system also, and as an inevitable side-effect, produced a large degree of supply variation between and within Palestinian communities. Compare, for instance, the quite different water situations in two small Palestinian villages near Hebron. Where the village of Duwarra lay fortuitously alongside the 16-inch Herodian-Kiryat Arba main line, and was thus in receipt of a constant supply of water even during the summer (the case of Duwarra will be discussed at greater length in subsequent chapters), the nearby village of Quasiba depended for its supplies on a 2-inch rubber tube running from the town of Sayyir (see Figure 3.3). Sayyir itself had neither the supplies nor the pumping capacity to supply Quasiba, with the result that Quasiba would go without piped supplies for five or more months each summer.[53] Equally, contrast the water situation at the lower end of Duwarra, near to its 2-inch connection with the Herodian-Kiryat Arba line, with that at the top of the village, where households faced perennial water shortages. Within virtually every Palestinian town and village in the southern West Bank, the highest houses would suffer much longer and more severe shortages than those below them – in part because the supplies received were largely insufficient to meet total demands; partly because pumping facilities within Palestinian distribution networks were generally inadequate or absent; and because, in combination of these two points, there was commonly insufficient pressure to pump water to the highest areas. Even more strikingly, contrast the situations in networked communities like Duwarra and Quasiba with those in the

58 southern West Bank Palestinian villages (and half of all West Bank villages) which were not linked to piped networks at all, and were thus forced to obtain all of their water supplies through other means.[54] In each of these regards it is clear that the West Bank's water infrastructure produced stark supply differences not just between Palestinians and Israelis, but also amongst the Palestinians themselves.

Not all of the Palestinians' water problems under occupation can be blamed on this techno-political apparatus, of course. In most Palestinian towns and villages, the internal water networks suffered from ideal – typically 'technical' shortcomings. Bethlehem's water network, for instance, was incredibly haphazardly designed: it featured supply bottlenecks (such as cases where 4-inch mains were supplied by pipes of only 2-inch diameter); had no supply loops (which in a normally functioning system maintain a constant circulation of water, thereby preventing both pressure build-ups, and deterioration in water quality); and lacked pressure control mechanisms (such that the lowest parts of neighbouring Beit Sahur, for instance, suffered not from perpetual shortages, but from an over-pressurised supply and consequently high levels of leakage).[55] Much of the piping was old and in poor condition. Even these technical deficiencies need to be understood, however, within the broad political economic context of the Israeli occupation – as a consequence of years of chronic under-investment in Palestinian services by an occupying power that was directing the major part of its resources to its own population.

This, broadly speaking, was the situation that West Bank Palestinians found themselves in at the onset of the Oslo process. In Tel Aviv, residents were in receipt of continuous water supplies, with water being conveyed from the Sea of Galilee and from springs and wells at the foot of the West Bank, and with most of their wastewater being treated and then transferred to the Negev for agricultural purposes, as discussed earlier. In the West Bank, by contrast, Palestinians were subject to the Military Orders and discriminatory regulations of the Israeli Civil Administration; their wells and distribution lines were owned and controlled by Mekorot; half of their villages had no access to piped water at all; those that did have internal networks found them in a state of general disrepair; they had little means of recycling their wastewater since treatment facilities were largely non-existent (domestic and industrial wastewater alike was simply transferred untreated, either through sewerage systems or by sewage tankers, to the nearest wadi); a disproportionate part of the region's water resources was being diverted to Israeli settlements (both because of the design of water networks, and as a result of the actions of Israeli and local Palestinian water administrators); they typically faced routine summer water supply shortfalls (though there were also wide variations in water supply between and within networked Palestinian communities); and they paid much more for their water than Israelis, let alone Israeli settlers. By one (Israeli) estimate, the average West Bank settler was in receipt of 12 times as much water as the average West Bank Palestinian.[56] These realities represented the sorry culmination of the long history of the Zionist movement's colonial encounter with the Palestinians. Could, though, this history be brought towards a close? Could Israel's apartheid policies be reversed? And what chance that the water situation in the Territories could be ameliorated? Having considered the historical backdrop to the Oslo period, it is to these more contemporary questions that we now turn.

PART TWO

The Oslo Period

CHAPTER 4

Dressing Up Domination as Co-operation

Most mainstream narratives of the course of the Oslo 'peace process' have characterised it according to what may be thought of as a 'breakthrough to breakdown' model.[1] Such narratives typically assume, first, that September 1993 marked a sharp discontinuity in Israeli–Palestinian relations, with the signing of the initial Oslo Accords opening 'a new era not only for the Middle East, but for the entire world'; second, that there was a series of further breakthroughs, most notably the Cairo and Oslo II Agreements of 1994 and 1995, and the Palestinian elections of 1996; third, that there exist significant policy and attitudinal differences between Labor and Likud administrations, such that under the latter the peace process has inexorably tended towards breakdown; and fourth, that the onset of the al-Aqsa intifada and the election of Ariel Sharon signalled the final dissolution and reversal of all that was achieved during the mid-1990s.[2] Commentators differ in their assessments of when the Oslo process collapsed, but few are in any real doubt that a critical breakdown, and possibly a terminal death, has now taken place.[3]

It is indeed hard to avoid the language of 'breakdown', 'collapse' and 'death' when analysing post-September 2000 events in Israel and the Occupied Territories – I have used such language myself at various points in this book. Nonetheless, the breakthrough to breakdown model represents but one way of reading the short history of the Oslo process. Many of the institutional and political changes that the Oslo process brought about were more cosmetic than real. Moreover, with the benefit of time, it may well be that the Oslo process is viewed less as a period of breakthrough between two distinct periods of violence (the intifadas of 1987–93 and 2000–?), than as a brief and relatively unimportant interlude within a single and increasingly militarised era of intifada-cum-war. Whether the signing of the initial Oslo Accords 'opened a new era for the Middle East' remains a moot point, and subject to historical reinterpretation.

To be sure, not all commentators have bought into the mainstream breakthrough to breakdown discourse on the Oslo process. In their critical accounts of the process, figures such as Noam Chomsky and Edward Said have questioned whether the Oslo process was ever really alive in the first place, and whether it ever constituted the enormous breakthrough that it was so often presented as being.[4] The argument developed in this chapter runs along broadly similar lines, contending with regard to water issues, but other ones besides, that the much-feted achievements of the Oslo process were often more cosmetic than real. The development of co-operative mechanisms for managing the West Bank's water resources, systems and supplies has often been lauded as one of the major successes of the Oslo process. I submit, rather, that routine co-operation between Israeli and Palestinian water managers was taking place long before the onset of the Oslo process, and that the distribution of powers and responsibilities between these Israeli and Palestinian water managers changed little between the pre-Oslo and Oslo periods. Much of what had previously been patron–client relations under occupation were suddenly discursively repackaged and re-presented as instances of Israeli–Palestinian 'co-operation'. Moreover, the problems that have beset 'co-operation' in the water arena since Oslo owe very little to policy differences between the Labor and Likud administrations. To the contrary, in terms of the control and management of the West Bank's water resources, systems and supplies, the continuities between the pre-Oslo, Oslo and breakdown periods are much more striking than the discontinuities between them. The main consequences of the Oslo water agreements, I argue, were not any significant transfer of power between Israelis and Palestinians, but rather three things: the construction of extra layers of bureaucracy which had few new powers, and which above all served to symbolise and dissimulate Palestinian autonomy; a transfer of power from Palestinian 'insiders' to PLO 'outsiders' returning from Tunis; and a transfer of some of the burdens of occupation from Israel to both the PA and the international donor community. With regard to Israeli–Palestinian relations, however, the Oslo process did little more in this particular sphere than to dress up domination as 'co-operation'.

I begin by providing an introductory sketch of the main consequences – achievements or otherwise – of the Oslo process. Thereafter, though, we concentrate solely on water issues. First we consider the water accords of the Oslo II Agreement, and some of the excitement to which these accords gave rise; then, in more critical vein, we dig beneath the surface of these apparently new co-operative mechanisms, discovering that all is not quite as it seems. It should be noted that I say nothing in this chapter about the arguments, negotiations or unequal power relations that lay behind these agreements. We consider these in chapter 6; here we focus solely on the substance of the agreements reached.

THE RESTRUCTURING OF OCCUPATION

Summarising all too briefly, it can be said that the Oslo Accords and process brought four main changes within the West Bank and Gaza: they led to the creation of the Palestinian Authority (PA), which had and still has many of the symbolic trappings of statehood but whose legislative, administrative, and territorial powers have always been heavily circumscribed; they granted the PA significant policing and security powers, such that it acted as 'Israel's enforcer' in the West Bank and Gaza; they brought about a limited restructuring of Israeli–Palestinian economic relations; and they licensed Israel's further colonisation of the West Bank, the settlement-building programme having continued unabated since 1993.[5]

Starting with the trappings of statehood, the Palestinians of the West Bank and Gaza now have, as a result of the Oslo process, an elected 'President' (though formally he is only allowed to call himself 'Chairman'), an elected Palestinian Legislative Council (the PLC), and a full range of ministries and agencies.[6] They have their own postage stamps (which are, however, subject to restrictions) and new PA identity cards (details of which are forwarded to Israel).[7] The PA is responsible for administering health, education and social welfare services to Palestinians throughout the West Bank and Gaza, for managing tourism, and for collecting direct taxes and VAT on local production.[8] And the PA also formally had its own autonomous territory. Following the May 1994 Cairo Agreement, Israeli military forces withdrew from around 80 per cent of the Gaza Strip and 4 per cent of the West Bank (around the small town of Jericho in the Jordan Valley), leaving them in the hands of the PA (see Figure 4.1).[9] Moreover, following the September 1995 Oslo II Agreement, the PA came to assume security responsibility within seven of the main West Bank Palestinian towns, as well as civil responsibilities within an additional

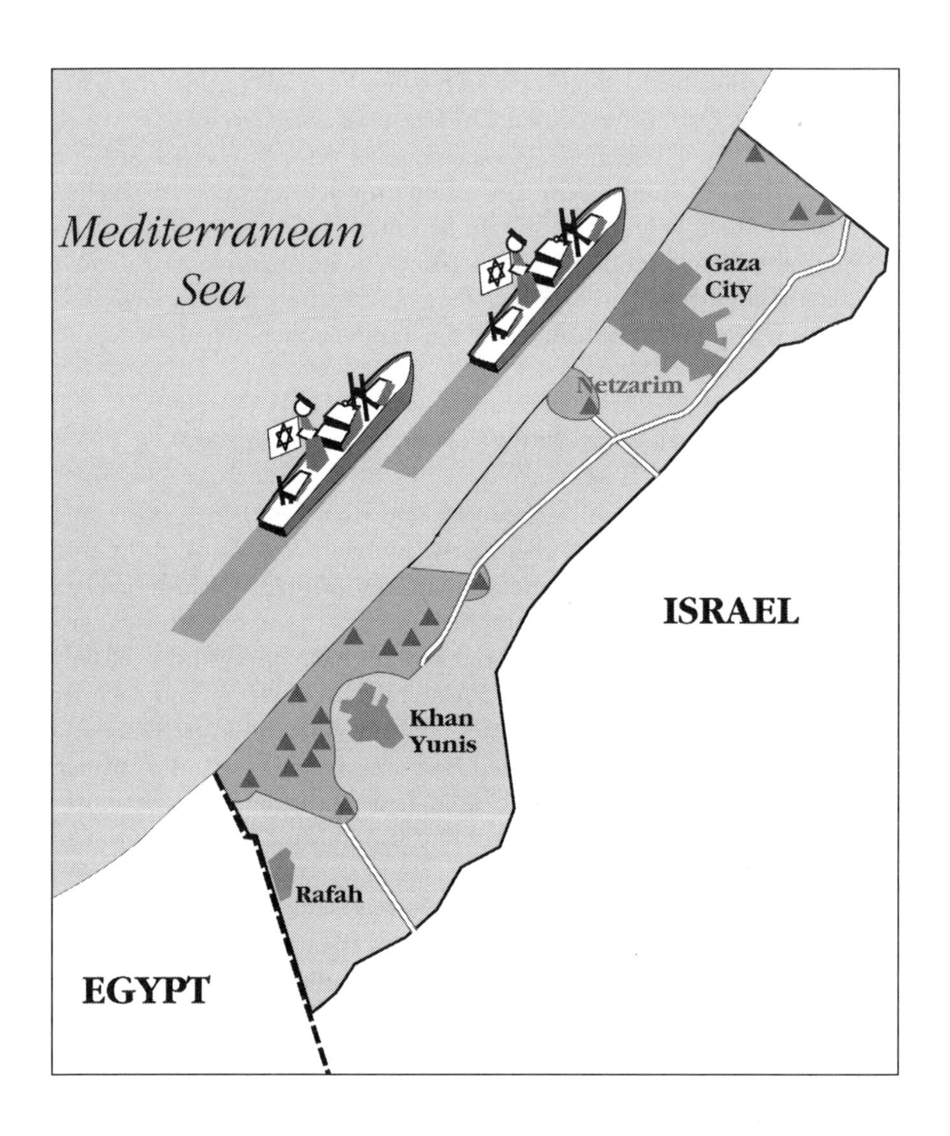

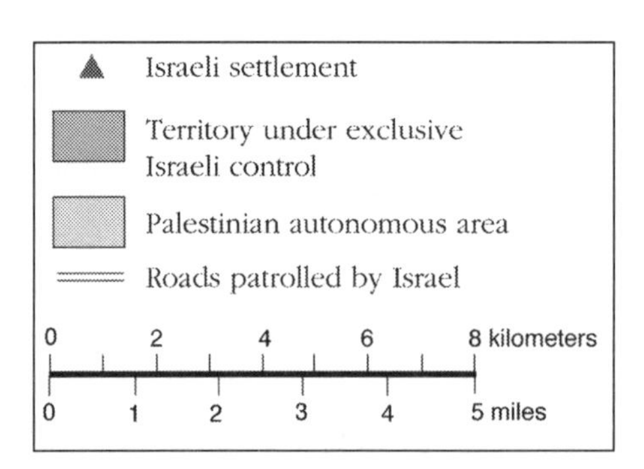

Figure 4.1 Palestinian Autonomous Area: Gaza Strip 1994 (reproduced with permission from Jan de Jong and the Foundation for Middle East Peace)

24 per cent of the West Bank, including most Palestinian towns and villages (see Figure 4.2).[10] These areas were added to as a result of further agreements negotiated in 1997, 1998 and 1999, such that by summer 2000 the PA had full security and civil control (Area A) over 17.2 per cent of the West Bank, and civil control (Area B) over an additional 23.8 per cent of the territory.[11]

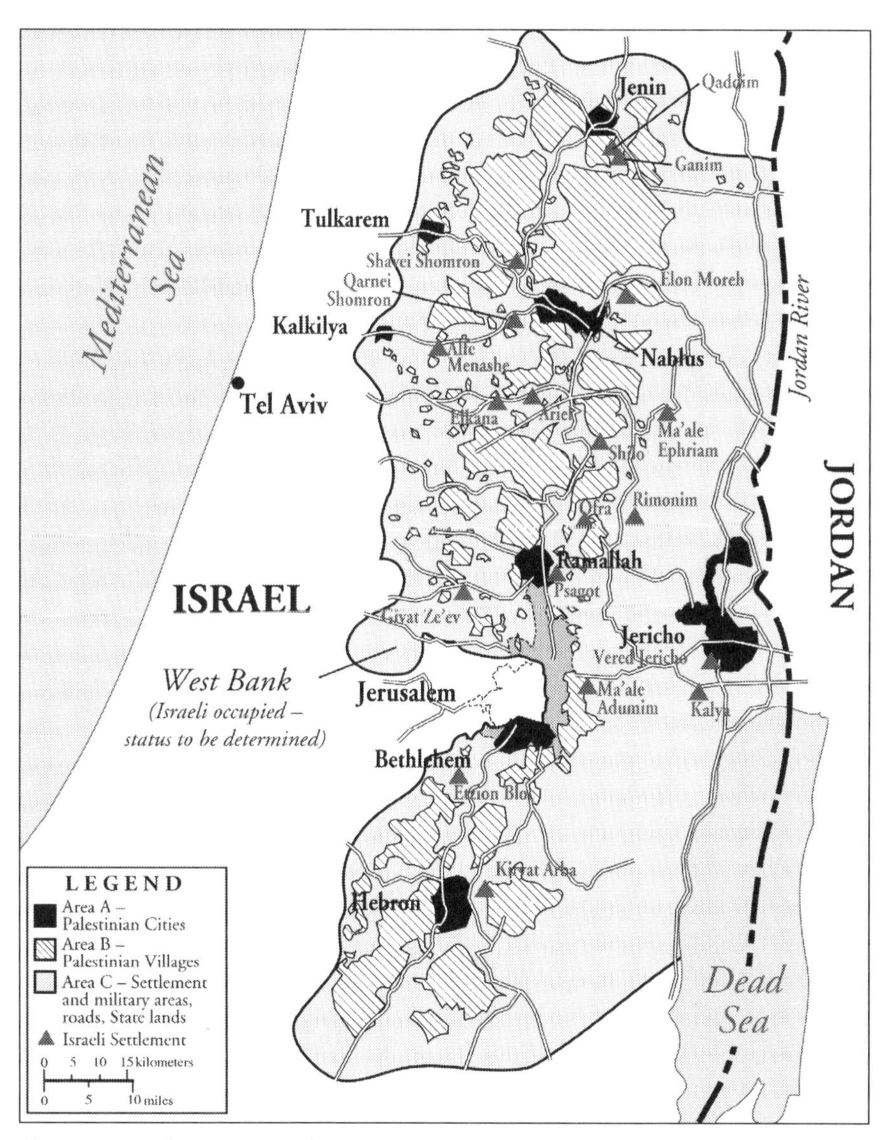

Figure 4.2 Oslo II Map Outlining Areas A, B and C (reproduced with permission from Jan de Jong and the Foundation for Middle East Peace)

These various territorial, administrative and legislative powers were in truth heavily circumscribed, however. By the terms of the Oslo Accords, the PLC is not and never has been entitled to amend or abrogate existing laws or military orders.[12] The PA has had no jurisdiction over Israelis, even when they are within autonomous Palestinian territory.[13] Most PA ministries and agencies have been overseen in their work by joint Israeli–Palestinian committees, which have often severely constrained their activities. Moreover, the territory formally controlled by the PA consists of scores of non-contiguous fragments – following Oslo II, there were 220 of these in total, 190 of them under 2 square km in size (see Figure 4.2).[14] A few of these cantons (Area A) formally come under 'full' Palestinian control, the majority however under a 'partial' control (Area B) that is limited to civil and not security affairs. Sixty per cent of the West Bank (Area C) and 20 per cent of the Gaza Strip remained throughout the Oslo process under full Israeli territorial jurisdiction. Since Operation Defensive Shield, of course, these territorial formalities have counted for very little.

The PA's most significant powers related to matters of policing and security. The Declaration of Principles had specified that the PA would have its own 'strong police force'; Oslo II stipulated that this force would number up to 24,000 officers.[15] By 1997, though, the PA's various police and security forces had perhaps 36,000 officers, such that the PA areas then had the highest proportion of police in relation to population in the world.[16] Besides the regular civil police force, as many as 14 intelligence and security services were operating in the West Bank and Gaza.[17] These security services did not legally exist, and contravened the terms of the Oslo II Agreement, which specified that the Palestinian police would 'consist of one integral unit under the control of the Council'.[18] They overlapped and competed in their work, and were all answerable to Arafat, giving him immense scope for patronage. Revealingly, Israel for the most part raised few objections to this inflation in the size and number of the PA police forces: neither Rabin nor Peres were particularly concerned about this matter, while Binyamin Netanyahu's objections constituted little more than a tactical attempt to defer international criticism of his government by arguing that the PA was likewise violating signed agreements.[19] Rather than seeking to rein in the PA police and security forces, Israel to the contrary gave the PA's intelligence and security services unofficial licence to operate across the West Bank and Gaza, even in the Israeli-controlled Area C. Under the terms of the officially non-existent January 1994 Rome Agreement, the PA became informally responsible for policing the Palestinian population across the West Bank and Gaza, this being in return for ongoing intelligence on the Palestinian and especially Islamist opposition. Significantly, this agreement was reached and became operational long before the PA was granted *de jure* control over its autonomous territories.[20] From even before the Oslo II Agreement, then, there was routine security co-ordination between the PA's security services and the IDF and GSS (Israel's General Security Services); since 1998, the CIA has also been formally in on the act.[21] In co-ordination with these US and Israeli agencies, the PA thus developed extensive internal security powers within the West Bank and Gaza, extending well beyond its *de jure* territorial enclaves; and it also, as it happened, managed to accrue a deplorable human rights record, with many of its abuses traceable to US and Israeli demands that it crack down on terrorists.[22] Yet irrespective of these powers of internal repression, Israel retained sole responsibility for external security, such that Israel controlled the international borders of the West Bank and Gaza, as well as movement between them. Israel was also free to impose 'internal closure' whenever it saw fit, effectively sealing off PA autonomous areas from one another and preventing people from leaving or entering them (from leaving or entering Hebron, for example). Israel never formally 'withdrew' from but instead 'redeployed' within the Territories.[23]

The Oslo process also inspired a limited restructuring of Israeli–Palestinian economic relations.[24] Ever since March 1993, the West Bank and Gaza have been subject to a general closure, with only limited numbers of Palestinians being permitted access into Israel, or to Jerusalem. Israel claims that this has pre-eminently been for security reasons; nonetheless the main result has been that the Palestinian day labour force employed within Israel has declined dramatically from its high of 165,000 in the late 1980s, often to zero.[25] This in turn has caused both high and fluctuating Palestinian unemployment levels, especially in the Gaza Strip, and also a labour deficit within Israel. Israel and Israeli business have responded to this in two ways: first, by importing workers from eastern Europe and southeast Asia; and second, by increasing the degree of subcontracting within the West Bank and Gaza, and establishing industrial zones in the vicinity of Palestinian population centres. This latter development would seem to suggest that the period since 1993 has involved not an end to, but a restructuring of, the relations of economic dependency established during the occupation – 'a transition from colonialism to neo-colonialism'.[26] In other respects, though, the dependency relations have barely changed: the West Bank and Gaza still constitute a captive market for Israeli produce, and most of the constraints imposed on Palestinian industrial and agricultural production during the occupation still remain in place, albeit now with PLO consent. Besides these developments in relations with Israel, the one major change resulting from the Oslo agreements was the emergence of the PA which, at its height employed an estimated 90–100,000 people in its police forces, schools and bloated ministries.[27] Yet here too we find dependency, both in relation to international donors and to Israel. With regard to the donors, throughout the Oslo period they were providing virtually all the finance for institution and infrastructure-building – and indeed have continued doing so since the collapse of Oslo, as we shall see in chapter 7. Israel, for its part, was under the terms of an early economic agreement with the PLO, forwarding 75 per cent of the income tax deducted from the wages of Palestinian day labourers in Israel to the PA, as well as 100 per cent of the income tax of those working in Israeli settlements.[28] These remittances – accounting under Oslo for around two-thirds of all PA revenues – have been withheld since December 2000, to disastrous effect.[29]

Finally, the Oslo process licensed and led to the extension of Israel's ongoing settlement programme in the West Bank and Gaza. Settlement construction continued unabated throughout the Oslo period, especially within the Greater Jerusalem area, and in strategically important areas between Palestinian population centres. By summer 2000 there were at least 80,000 more settlers in the West Bank alone (excluding East Jerusalem) than there were in 1993.[30] Also under Oslo, countless new roads were constructed across the West Bank, these connecting the expanding settlements with Israel, and enabling settlers to bypass the autonomous Palestinian enclaves.[31] Besides serving a territorial function, these settlements and bypass roads have also furthered the Israeli state's capacity to police and contain the Palestinian population: they have effectively separated most of the major population centres from one another, and have rendered it possible for the IDF to enforce internal closures without overly disrupting the lives of its settlers. These developments continued relentlessly between 1993 and 2000, irrespective of whether Rabin, Peres, Netanyahu or Barak was at the helm. In the lead-up to the July 2000 Camp David negotiations, for instance, with Ehud Barak and the Labor Party in power, the rate of settlement construction increased by a staggering 81 per cent.[32] And while the Palestinians complained throughout the Oslo period about Israel's 'unilateral actions', nothing in the Oslo Accords precluded the Israeli state from extending its territorial reach over Area C of the West Bank, or over the Jerusalem area, as it saw fit.

THE OSLO II BREAKTHROUGH ON WATER ISSUES

With this broad context established, we can now turn to the water arena. Of the water agreements reached between Israel and the PLO during the course of the Oslo process, by far the most important are to be found within the September 1995 Oslo II Agreement. The 1993 Declaration of Principles had said very little regarding water issues, calling for the creation of a 'Palestinian Water Administration Authority', and for '[c]o-operation in the field of water, including a Water Development Program prepared by experts from both sides, which will specify the mode of co-operation in the management of water resources in the West Bank and Gaza Strip, and will include proposals for studies and plans on the water rights of each party, as well as on the equitable utilization of joint water resources for implementation in and beyond the interim period' – but beyond these vague commitments the Declaration had barely mentioned the issue.[33] Building upon this, the 1994 Cairo Agreement had stipulated that, with the exception of water supplies to Israeli settlements and military areas, all water resources and systems in Gaza and the Jericho Area would be 'operated, managed and developed' by the PA.[34] Neither Gaza nor the Jericho Area, however, are home to abundant water resources: both of them are downstream areas with shallow and highly saline underground reserves, ones that, if mismanaged by the PA, could do little to endanger Israeli water supplies. The Cairo articles on water hardly betrayed evidence of Israeli generosity. By contrast, the water accords of the Oslo II Agreement seemingly paved the way for the joint Israeli–Palestinian management of the West Bank's rich underground water resources.

Oslo II contained the first explicit and unequivocal recognition of 'Palestinian water rights in the West Bank', precise details of which would be agreed upon during permanent status negotiations.[35] More significantly (at least in the short term), Oslo II committed Israel and the PA to establishing a 'Joint Water Committee' (JWC), with responsibility for overseeing the management of all of the West Bank's water and sewage resources and systems.[36] The JWC would operate in seemingly egalitarian fashion: it would be made up of an equal number of Israeli and Palestinian representatives, and decisions within it would be reached by consensus.[37] The JWC would have overall authority for surveying and protecting existing resources, for developing supplies, for maintaining existing infrastructures, and for constructing new ones.[38] The JWC would not, however, be responsible for the day-to-day management of resources and systems; it would function, rather, as a co-ordinating body, with most on-the-ground work being undertaken separately by one or other of the parties. Thus particular water and sewage systems would be controlled by either Israel or the PA: those systems 'related solely to Palestinians' which, until then, were 'held by the military government and Civil Administration' would be transferred to the PA, while all other systems would remain under Israeli control.[39] Israeli and Palestinian water authorities would operate separately, but under the overall control and direction of the JWC. Irrespective of its name, the JWC would be a 'co-ordinated' and not a 'joint' management structure.[40]

Oslo II also stipulated that the two sides would establish, under the supervision of the JWC, 'no less than five Joint Supervision and Enforcement Teams (JSETs)' for the monitoring and policing of the West Bank's water resources, systems and supplies.[41] As with the JWC itself, the JSETs would operate according to strictly egalitarian principles: each of them would be comprised of 'no less than two representatives from each side', and each side would have its own vehicle and cover its own expenses.[42] The JSETs teams would be responsible for locating unauthorised water connections, for supervising infrastructure developments, and for monitoring well extractions, spring discharges and water quality.[43]

Such management duties aside, one of the major and immediate tasks of the JWC would be to oversee the development of additional waters for the West Bank's Palestinian communities. Oslo II committed Israel and the PA, between them, to developing during the interim period 23.6 mcmy of water from the West Bank's underground aquifers, 'in order to meet the immediate needs of the Palestinians'.[44] In addition to this, Oslo II defined 'the future needs of the Palestinians' at an additional 70 to 80 mcmy.[45] To put these figures in perspective, it is worth noting that, as of 1995, total water use amongst West Bank Palestinians officially stood at a mere 118 mcmy.[46] The clear promise of Oslo II was that the West Bank's Palestinian communities would soon be receiving significant new and additional quantities of water.

These terms have routinely been praised as amongst the most significant of the Oslo II Agreement. The Israeli press lauded the agreement on water rights as a 'breakthrough'.[47] Some observers claimed – with hyperbole that is unfortunately all too frequent when it comes to discussion of Middle Eastern water issues – that the Oslo II water accords constituted the most significant result to date of the entire Oslo process.[48] Others, more judiciously, ventured that the 'water provisions of the Interim Agreement represent a major step towards a permanent Israeli–Palestinian accommodation over water', a 'step in the direction of an equitable water-sharing arrangement'.[49] Not all agreed, of course. From the Israeli right, the Oslo II terms were denounced as 'a giveaway of our water to the Arabs'.[50] Seen in this light, the agreement irrevocably effected a 'loss of control over a major part of the country's natural water sources to Arab authorities'; hence 'for Israel,' claimed Martin Sherman, 'the hydro-political future in the wake of the Oslo Accords appears both bleak and risk-fraught'.[51] Yet critical as these right-wing Israeli voices were, they nonetheless depicted the Oslo II terms on water as marking, if not a positive breakthrough, then at least a sharp discontinuity in the management and control of the West Bank's waters. International observers and Israeli critics alike generally perceived these water provisions as 'opening a new era' in Israeli–Palestinian water relations.

Most commentators have likewise evinced stark differences between Israel's various Likud and Labor administrations in their attitudes towards water co-operation with the Palestinians. Commentators have suggested that the 'transition' promised by Oslo II thereafter gave way to 'stalemate' such that, after 1997, co-operation within the JWC largely came to a halt.[52] PA water officials have argued along similar lines that Netanyahu's Likud administration 'continuously attempted to destroy the agreements and destroy water projects', and furthermore, that 'there was nothing wrong with the agreement', the problems lying instead in the interpretation and implementation of this agreement by a hostile Israeli government.[53] For some, the election of Ehud Barak in May 1999 and the subsequent agreement of the Sharm-El-Sheikh Memo raised new 'hope and optimism' that outstanding water issues might be resolved.[54] In one way or another, all such commentators have conformed to a breakthrough to breakdown narrative of the peace process as a whole, and of Israeli–Palestinian water co-operation in particular.

There have admittedly been some exceptions to this general rule. Many Palestinian water experts, for instance, criticised the Oslo II Agreement for its deferral of water rights questions to final status negotiations, as well as for its tacit legitimation of Israeli access to 'Palestinian water resources' for the duration of the interim period.[55] Implicit in such criticisms is the claim that the Oslo II water provisions did not constitute a significant breakthrough, and to the extent that Oslo II did not address water rights issues, and granted Israeli settlers continuing access to West Bank water supplies, these criticisms are surely valid (under the terms of Oslo II, Israel would continue to consume 87 per cent of the total water yield of the West Bank's two trans-boundary aquifers for the duration of the interim period, with Palestinians consuming a mere 13 per cent of these waters).[56]

Nonetheless, such criticisms do not go quite far enough, since they only criticise the joint management mechanisms put in place by Oslo II, without questioning whether these new mechanisms were really as novel as they may at first have seemed. As we will see below, much of what was agreed in the Oslo II negotiations did little more than formalise and legitimise management structures and relations which were already very much in existence.

AUTONOMY IN THE WATER SECTOR?

Israel's recognition of Palestinian water rights aside, the Oslo II water accords achieved three main things: they inaugurated a formal system for the co-ordinated management of the West Bank's water resources, systems and supplies; they established a formal system of teams (the JSETs) for supervising and monitoring these resources, systems and supplies; and they stipulated that additional water would be made available to the West Bank's Palestinian communities. But precisely how new – and how significant a breakthrough – were each of these apparent achievements?

Managing Resources, Systems and Supplies

During the course of the occupation, Israel had constructed both an integrated water supply network across the West Bank and a delegated institutional regime for managing the Palestinian water sector, as described in the previous chapter. Its water supply system – a complex apparatus of pipework and politics – conjoined Israeli settlements and Palestinian towns and villages within a single web, but simultaneously discriminated between them. Its institutional regime, meanwhile, was premised on the Palestinian-staffed West Bank Water Department, and Palestinian municipalities and village councils, being responsible for liaising with Palestinian water users. While the Israeli Military Government (later Civil Administration) and its Water Officer retained overall regulatory control, and while Mekorot owned the water supply infrastructures, it was nonetheless the Water Department and local Palestinian authorities which were responsible for maintaining distribution lines, for opening and closing supply valves to Palestinian communities, and for billing Palestinian communities. None of these Palestinian institutions had any power over or responsibility for Israeli settlers, however; the Water Department was not allowed to close water supply valves to Israeli settlements, for instance, and had no role in billing Israeli settlers. These Palestinian institutions, and the Water Department in particular, thus functioned as key institutional interfaces between the Military Government and the occupied Palestinian population, enabling the Israeli state to effect its colonial and apartheid water policies without having any direct contact with Palestinian users.

As has already been noted, the water accords of the Oslo II Agreement set in place mechanisms for the co-ordinated management of the West Bank's water resources, systems and supplies. A Joint Water Committee would be established to oversee this co-ordinated management system. Supply infrastructures, however, would not be managed directly by the JWC, but by one or other of the parties; systems 'related solely to Palestinians' would be 'operated and maintained by the Palestinian side solely', while all other systems would remain under Israeli control. The implications of this should by now be readily apparent. The Palestinians would henceforth be responsible for maintaining and operating internal systems within Palestinian towns and villages, as well as those connections to such internal systems which did not feed Israeli settlements. Yet given that by 1995 Israeli and Palestinian water supply networks were thoroughly integrated, this did not promise the Palestinians a great deal. Israel would continue to control the vast majority of supply lines, and would also continue to control the numerous deep wells which had been drilled by Mekorot since 1982, since these all supplied at least some Israeli settlements. Moreover, given that most local water supply and infrastructure management within the West Bank was already being undertaken by Palestinians – both by the West Bank Water Department, and by municipalities and village councils – the seeming novelty of Oslo II's co-ordinated management system was largely illusory. Hence Dellapenna is only half right when he observes that the Oslo II water accords did but 'reinforce the dependence of the Palestinians on Israeli water facilities, in effect converting Israelis into the "upstream" partner in developing from the aquifer'.[57] It is true, as he suggests, that under the terms of Oslo II, Israel retained control over the West Bank's water resources, while the Palestinians were granted responsibilities only in the management of local water supplies. It is mistaken, however, to suggest that Oslo II in any way transformed, reconfigured or 'converted' the responsibilities of either Israelis or Palestinians in managing these supplies. To the contrary, the water accords of the Oslo II Agreement merely formalised a supply management system which had been in operation for years, presenting it, misleadingly, as part of an egalitarian-sounding 'joint' and 'co-ordinated' management system.

Very much the same can be said regarding water prices. Oslo II stipulated that 'in the case of purchase of water by one side from the other, the purchaser shall pay the full real cost incurred by the supplier, including the cost of production at the source and the conveyance all the way to the point of delivery'.[58] At first glance this would appear fair and reasonable. As noted above, however, the Israeli authorities would continue to exercise control over the West Bank's water resources, and over all 'upstream' facilities, such that the Israeli authorities would always be the 'suppliers', Palestinian authorities and communities the 'purchasers'. Moreover, the terms of this article apply only to transactions between Israelis and Palestinians, placing no constraints on purchases by Israeli settlers. As we saw in the previous chapter, the latter pay for their water at highly subsidised rates. Thus under the reasonable-sounding terms of Oslo II, Palestinians would have no option but to pay the 'full real cost' of production and supply to the Israeli authorities, while these same authorities would be free to continue supplying settlers at rates well below the real cost of production and supply. As with the management of systems and supplies, Oslo II simply legitimised a discriminatory pricing mechanism which had existed well before 1995.

Beyond this, the Oslo II arrangements had one extra benefit for Israel. Since the onset of the intifada in 1987, the West Bank Water Department had been facing increasing levels of non-payment by Palestinian municipalities and individuals, such that by 1995 it had debts of around NIS 18 million ($4.5 million). With the inauguration of a formal 'joint management' system, these debts suddenly became taken on by the Palestinian side, being covered by the Palestinian Ministry of Finance. By 2002, these Water Department debts had risen to NIS 110 million ($24 million).[59] The formalisation of Israeli–Palestinian co-operation had enabled Israel to divest itself of some of the most onerous burdens of occupation, without losing control of either water resources or supplies to Israeli settlements, and without having to forego its discriminatory pricing policy.

Monitoring Resources

If we turn now to consider the Joint Supervision and Enforcement Teams (JSETs), we find something conspicuously similar. Besides maintaining the West Bank's water networks and billing Palestinian customers, one of the Water Department's main tasks under occupation involved monitoring the West Bank's springs and wells. In this, Palestinian technicians within the Water Department followed a system developed during the late 1960s and early 1970s under the direction of the Israeli Hydrological Service (IHS).[60] As early as September 1967, the IHS and Water Department had begun developing procedures for the monitoring of the territory's water resources. During the first few years of occupation, the two institutions together measured and classified all the territory's springs and wells. Springs were categorised according to discharge, and hence according to the frequency with which they were to be monitored; certain representative wells were selected for routine monitoring; and schedules and data collection techniques were all established and standardised. By the early 1970s, a hydrological monitoring system was in full operation. Monitoring was for a time conducted jointly by Israeli and Palestinian technicians, but thereafter only by the latter.[61]

Oslo II stipulated that 'no less than five' JSETs would be established under the direction of the JWC to monitor and police the West Bank's water resources, systems and supplies.[62] Three such teams were immediately established, each of them responsible for hydrological monitoring.[63] The twist here lies, though, in the fact that these three JSETs followed precisely the same monitoring system as had been followed since the early 1970s by the West Bank Water Department. Monitoring was conducted by the same Palestinian technicians, and in line with the very same procedures and schedule; and data was recorded on forms which barely differed from those which had been used prior to the Oslo II Agreement (compare the two forms shown together as Figure 4.3). Formally speaking, a system of 'joint supervision' and co-operation had replaced one of occupation and domination, but in terms of the monitoring work which was actually undertaken, changes were only minimal and by no means altogether positive for the Palestinians.

The new JSETs regime brought about three main changes, none of which are as significant or as beneficial as they may initially appear. First and most obviously, following Oslo II the Water Department's workers were now accompanied on site visits by Israelis. Nonetheless, the Palestinians remained the ones conducting hydrological readings, with the Israeli teams 'just writing down the numbers'.[64] Hence in this regard, the new JSETs system did little more than to return monitoring procedures to those of the early 1970s, when Israeli technicians accompanied their Palestinian counterparts around the West Bank's water sources. Second, whereas under occupation Palestinian technicians would carry out their work without any escort, JSETs teams have always

STATE OF ISRAEL
MINISTRY OF AGRICULTURE
WATER COMMISSION
HYDROLOGICAL SERVICE

COMPUTATION OF DISCHARGE MEASURED IN SPRINGS BY MEANS OF CURRENT-METER

District __________ No. of means. CM __________

Name of site __________ Coordinates East ________ North ________

Date Hour	Current Meter Type & No.	Last Calibr	Ring every	Stop Watch No.	Weather	Wind With Against Flow None	Colour of Water

Bed Covered With	Vegetation In Channel	Type of Flow
Large Stones/Small	Shrubs	Quiet
Stones/Pebbles	Grass	Turbulent
Sand/Mud/Concrete	Clean	Disturbed

Water Sample Taken / Not Taken

Remarks __________

Result of the present measurement Q ________ m^3/ Sec.	Result of the last Measurement
Add or Deduct:	Date ______ No. ______
	________ m^3/Sec.
Total ________ m^3/Sec.	

Sign. of observer ________ Sign of Computer ________

Date __________

Sign of Distr. Eng. __________

Date __________

Dist. from Initial Point In M.	Depth. of Water In M.	Instr. Below Surface M.	No. of Revolutions N.	Time In Sec.	Rev/ Sec.	Water Velocity M/Sec At Instr. M^3/ Sec. V	Mean In Vertic Vm	Mean Betw. Two Vertic Vm	Area of Wetted cross Section M^2	Vm.A In M^3/ Sec.

Cross Sectional Area ________ M^2 Q (M^3/Sec.)

Appendix 3H(2)

Joint Water Committee

J.S.E.T - Joint Supervision and Enforcement Team - Israeli-Palestinian

Computation of Discharge Measured in Springs by Means of Current-Meter

Form for current-meter measurement of springs

District: __________ No. of means. cm. __________

Name of site: __________

Date Hour	Current Meter Type & No.	Last Calibr	Ring every	Watch No.	Stop	Weather Wind With Against Flow None	Colour of Water

Bed Covered with	Vegetation in Channel	Type of Flow
Large Stones/Small	Shrubs	Quiet
Stones/Pebbles	Grass	Turbulent
Sand/Mud/Concrete	Clean	Disturbed

Water Sample Taken / Not Taken

Remarks __________

Result of the present measurement	Result of the last measurement
Q ________ m^3/sec.	Date: ________ No.: ________
Add or Deduct:	________ m^3/sec.
Total: ______ m^3/sec.	

Signature of Observer __________

Date: __________

Signature of District Engineer: __________

Date: __________

Signatures

Israeli Team	Palestinian Team
Name: __________	Name: __________
Signature: __________	Signature: __________

Distance from Initial Point in M.	Depth of Water in M.	Instr. Below Surface M.	No. of Revolution N.	Time in sec.	Rev/ sec.	Water Velocity m/sec. At Instr. m^3/sec V	Mean in Vertic Vm	Mean betw. Two Vertic Vm	Area of Wetted Cross Section M^2	Vm.A in m^3/sec

Cross Sectional Area ________ M^2 Q(M^3/Sec.)

Figure 4.3 Monitoring Under Occupation and Autonomy (made available by permission of West Bank Water Development)

been accompanied by Israeli soldiers, and sometimes also by Palestinian police. Yet here too an important rider must be offered: given that the West Bank is so fragmented between these zones, and given also that adjacent districts are under the control of different military and police officers, the task of organising security convoys unsurprisingly causes immense logistical difficulties, and consumes a large amount of time.[65] According to the JSETs' Palestinian co-ordinator, monitoring under Oslo took 'double the time' that it did under occupation.[66] In each of these regards, the main achievement of the JSETs regime was simply to create an extra layer of bureaucracy and a great deal of additional labour for Israeli and Palestinian water managers.

Third, following the agreement the Palestinians became entitled to make use of JSETs data. Under occupation, the Water Department had no means of aggregating and abstracting data, and hence record sheets were simply stored in the Water Department office by Beit El, copies being collected once a month by someone from the IHS.[67] By contrast, following the Oslo II Agreement the Water Department began forwarding copies of its records to the Palestinian Water Authority (PWA) which in 1996, and with the support of various international donors, started developing its own water resource databases.[68] Both Israeli and Palestinian water authorities now have access to JSETs data. This is evidently significant, and might well be taken as grounds for characterising the new JSETs system as a prime example of truly 'joint management'. Nonetheless, the PWA was utterly dependent on international donors in developing its water databases – and donors showed little interest in funding such low profile work, being much keener to 'fly the flag' over highly visible and prestige projects.[69] The PWA, in addition, had little access to past hydrological data, and neither did it have access to some of the most important current data, since the Israeli authorities consistently refused to transfer key information on extraction levels from wells located within Israeli settlements.[70] Given this, the PWA's water databases are heavily incomplete and of little practical utility, such that Palestinian water planners and negotiators and international donors alike have remained wholly reliant on Israeli databases, plans and models.[71] Israel has traditionally kept tight and secretive reins over its most important water-related information, and has arguably used this info-control to its advantage in negotiations with the Palestinians (as we shall see in the next chapter).[72] While under Oslo II the PA was granted the opportunity to use Palestinian-collected data – no longer did the Palestinians simply transfer record sheets to the Israeli authorities – the PA was nonetheless denied the opportunity of making meaningful use of this information. There may well have existed a formal mechanism for the joint supervision of the West Bank's water resources, but it was one which continued to enshrine overall Israeli control over water-related information.

Developing New Supplies

So much one might be willing to concede; but didn't the Oslo II Agreement also hold out the promise of additional water supplies for the West Bank's Palestinian communities? Indeed it did: 23.6 mcmy would be made available within the West Bank in order to meet the 'immediate needs of the Palestinians ... during the interim period', while a further 41.4–51.4 mcmy would be developed to meet the 'future needs' of West Bank Palestinian communities.[73] Yet significant as these provisions undoubtedly are, their overall import is qualified in a number of regards.

In the first place, these provisions placed only a minimal burden on Israel. Of the total promised new and additional supply to the West Bank of 65–75 mcmy, Israel would be financially responsible only for the development of 4.5 mcmy, with the Palestinians bearing the capital costs of developing the remaining 61.5–71.5 mcmy. Moreover, Israel would have to sacrifice only a minimal volume of water since, of the planned additional West Bank supply of 65–75 mcmy, Israel would only have to supply 3.1 mcmy from its national water system.[74] In these respects, the Oslo II Agreement simply enabled Israel to divest itself of the burden of developing much-needed additional waters for the Palestinians, transferring the financial burden for improving Palestinian water supplies from Mekorot to the international donor community and in turn the PA (which will at some point have to start repaying its soft loans to international donors).

All of the water not made available by Israel from its national water network would be developed 'from the Eastern Aquifer and other agreed sources in the West Bank'.[75] The Eastern Aquifer was named here in particular because, according to Israeli-derived hydrological data included alongside the Oslo II water accords, this was the only one of the West Bank's three underground bodies of water which was not yet being exploited to its fullest. By happy coincidence, its additional potential yield – estimated in Oslo II as 78 mcmy – would be just sufficient to meet all of the Palestinians' immediate and future water needs.[76] However, there is compelling evidence that this vastly overstates the remaining potential of the Eastern Aquifer. Water table levels are already rapidly declining in parts of the aquifer (the water level in one of the wells, Herodian 3, dropped by 85 m between 1981 and 1997); much of the aquifer's waters are highly saline, and would possibly need to be desalinated at great expense if they were to be used for domestic or agricultural purposes; most startlingly of all, one of the Israeli hydrologists who produced the figure of 78 mcmy for the Oslo II Agreement discounts the possibility that its entirety could ever be exploited on a sustainable basis.[77] The PWA and international donors have started developing new supplies from the Eastern Aquifer, its first new waters having come on tap in late 1999.[78] Nonetheless, the remaining potential of the aquifer is far below that officially given in the Oslo II Agreement. Thus by way of a second qualification, it can be said that the newly granted Palestinian right to further develop the Eastern Aquifer – seemingly an act of great Israeli generosity – is unlikely to yield its expected and hoped-for benefits.

As a third qualification, the structure of the JWC also serves to set constraints on Palestinian development of the West Bank's water resources. We have already seen that decisions within the JWC operate by consensus. Yet given that all infrastructure development works 'require the prior approval of the JWC' (and this includes every pipeline of greater than 2-inch diameter or 200 m in length, and includes every well that needs constructing or rehabilitating), it so follows that each of the parties has an effective veto over the other's proposals.[79] While in principle this applies equally to both sides, in practice it places by far the biggest constraints on the Palestinians, simply because they are so much more needful of new and additional supplies. As it has turned out, Israel has generally vetoed the Palestinian development of 'other agreed sources in the West Bank'.[80] It has rejected several proposed well locations on the grounds of them being too close to Israeli settlements.[81] Moreover, the PA has only succeeded in avoiding the Israeli veto on its infrastructure development proposals by entering into a tacit *modus vivendi* with the Israeli authorities, one in which Israel has been willing to grant licenses for Palestinian development of the Eastern Aquifer, but only in return for permission to construct new and enlarged water supply systems from within the Green Line to Israeli settlements in the West Bank (the Oslo II Agreement places no limit on new supplies to Israeli settlements).[82] While the PA has assented to this new construction work only on condition that it is not taken as implying recognition or acceptance of Israeli settlements (letters apparently passed between Jamil Tarifi, PA Minister of Civil Affairs, and the Israeli Ministry of Defence to this effect), the fact remains that the PA has in practice had little option, under the seemingly egalitarian terms of Oslo II, but to assent to the extension and entrenchment of Israeli 'facts on the ground' throughout the West Bank.[83]

As if this were not bad enough, the PA is not entitled to unilaterally amend or abrogate any of the laws or military orders which were in place on the eve of the Oslo II Agreement, the consequence of this being that all those water-related military orders which were put in place by the Israeli authorities in the wake of the 1967 war remained in force after Oslo II.[84] Ultimate decision-making authority over water resources and systems continued to lie with the Water Officer of the Civil Administration, who could in theory veto any Palestinian infrastructure development proposal, even after it has received the consent of the JWC. Such in fact has occurred on numerous occasions within the Israeli-controlled Area C, especially when proposed well locations and supply lines clashed with Israeli plans for new settlements and bypass roads.[85]

In each of these four regards – the facts that Palestinians and international donors carried almost all the responsibility for developing new supplies; that the Eastern Aquifer has a much smaller remaining potential than is officially recognised; that the structure of the JWC places greater constraints on the Palestinians than on Israel; and that the Civil Administration still retained an ultimate veto over Palestinian water developments – in each of these regards, the promises of new and additional supplies contained in the Oslo II Agreement were of much less significance than at first appears. Each of these four qualifications, it should be emphasised, follow directly from the terms of the Oslo II Agreement, not from their *post hoc* interpretation and implementation. There has indeed been little clear correlation between the state of Israeli–Palestinian water relations on the one hand, and the presence of Labor or Likud governments in Israel on the other. There have been significant delays in the approval of projects irrespective of whether Labor or Likud have been in power. Admittedly there was a relative breakdown in Israeli–Palestinian water relations during 1997, with Netanyahu at the helm in Israel; but contrast that with the fact that the *modus vivendi* detailed above emerged during 1998, also during Netanyahu's tenure.[86] Under Netanyahu, perhaps the main difference lay in the realm of rhetoric, with Israeli officials, including Ariel Sharon and Environment Minister Rafael Eitan, going so far as to accuse the PA of waging a premeditated 'sewage intifada' against Israel.[87]

Unsurprisingly, not all Israeli and Palestinian actions in the water sphere have operated within the terms of Oslo II. On the Israeli side, pipelines have on several occasions been laid to West Bank settlements without having first received JWC permission (and in some cases where Israeli proposals have been rejected by Palestinian JWC officials).[88] In cases where the Israeli authorities cannot achieve their projects through the legal-institutional mechanisms of the JWC, they can always resort to their far superior coercive capabilities to ensure that their pipelines get constructed as and when they require. Take, for instance, the words of Taher Nassereddin of the Water Department in recalling a Palestinian attempt to implement a sewerage project in the town of Salfit: 'We got the approval of the JWC one year ago. Suddenly a week ago they stopped the project, and the army went and took away the equipment. Why? Because they didn't take the permission of the officer in charge of water affairs because this is [Israeli-controlled] Area C. The donors they were surprised and astonished... They have the army, they have the force, we don't have. I know many [Israeli] projects were executed without [the permission of] the JWC.'[89] This hardly represents a model of joint and co-ordinated management.

DRESSING UP DOMINATION

The Oslo II Agreement undoubtedly engendered some important institutional and material changes in the management and development of the West Bank's water resources, systems and supplies. Two such changes stand out above all. The agreement inspired, most importantly, a massive influx of development aid, with money being channelled into the rehabilitation and construction of supply systems, as well as into the creation of the PA's water institutions. Problem-ridden though it has often been, this development aid had nonetheless brought improvements in the regularity and quantity of water supplies to many of the West Bank's towns and villages, as well as scattered improvements in the collection and treatment of wastewater and sewerage (these issues will be addressed further in chapter 6).

Beyond this, the Oslo II Agreement also resulted in the creation of new institutional arrangements, and a new distribution of decision-making powers, on the Palestinian side. Prior to Oslo II, the West Bank Water Department was the key Palestinian water institution, acting as an interface between the Israeli military and water authorities on the one hand, and Palestinian municipalities, village councils and individuals on the other. After Oslo II, institutional arrangements became a great deal more complex (as well as bureaucratised). The Palestinian water sector is now formally under the authority and purview of the PWA.[90] In many respects, however, the PWA is little more than a donor construct, its main responsibility being to co-ordinate donor projects, and the vast majority of its personnel being employed on a project basis (as of summer 2002, only five of the PWA's West Bank and Gaza personnel were on the PA payroll).[91] Moreover, as during the occupation, the West Bank Water Department is in many respects the most important Palestinian water institution within the West Bank, as the Water Department, and not the PWA, undertakes everyday water management. Simply put, the PWA oversees projects, while the Water Department undertakes mundane water management, very much as it did under occupation. The Water Department is still officially part of the Civil Administration. Thus the main institutional change within the Palestinian water sector has been the creation of a new and financially well-endowed top tier of administration, one that is defined and exists through its relations with the international donor community. As within so many parts of the PA, the very top of this new top tier is headed by two 'outsiders' (Nabil Sharif, head of the PWA, and Fadel Qawash, his deputy), who returned to the Territories from Tunis shortly after the onset of the Oslo process, and who are closely associated with and loyal to Yasser Arafat.[92]

In other respects, the changes effected by the Oslo II Agreement have been predominantly discursive rather than material or institutional. We have seen above that many of the Water Department's patron–client responsibilities under occupation were simply repackaged and re-codified by the Oslo II Agreement as elements of a 'co-ordinated' management system. Israeli and Palestinian water managers are evidently keen to obscure this fact, partly because this raises difficult questions about the significance of the peace process, and partly, in the Palestinian case, out of an understandable desire not to say too much about the key roles that fellow Palestinians played in administering water under the occupation, and to a degree facilitating it. Yet allusions to the less-than-substantive changes wrought by the Oslo process do occasionally seep out. Discussing the Oslo II negotiations and agreements – or what he woefully misrepresents as the 'dismantling of occupation' – Israeli negotiator Uri Savir observes with surprising candour that 'I feared that what would emerge from this makeover was more of the same on different stationery'.[93] This admission provides a curiously apt diagnosis of many of the consequences, often more discursive than substantial, of the water articles of the Oslo II Agreement. Israeli–Palestinian hydro-political relations within the West Bank were suddenly presented as 'co-operative' – rather than, say, 'oppressive', which is how we would surely characterise these relations as they existed during the occupation – not because certain management or monitoring procedures had changed, but because the signing of the Oslo II Agreement bestowed on these procedures a newfound legitimacy. 'Co-operation', within this agreement, denotes not a practical and material set of relations which are the antithesis of 'domination', but a discursive condition which arises and exists on the strength of a single tacit rule, one that stipulates that co-operation only occurs between free and equal consenting parties. Israel and the PLO signed the Oslo II Agreement as juridically free and equal parties – this, of course, being irrespective of the fact that one of the parties was vastly less free than the other, and that the parties were far from equal in their actual (military, political, institutional and economic) capabilities – and it was through this legal act that Israeli–Palestinian water relations within the West Bank became re-presented as 'co-operative'.

One might perhaps counterargue that the above ignores the fact that Oslo II was intended as a transitional arrangement, one that was not necessarily absolutely just, but which nonetheless represented 'a step in the direction of an equitable water-sharing arrangement'.[94] I would disagree. It would be a mistake to evaluate the merits or otherwise of the Oslo II Agreement on the grounds of the avowed intention of exchanging occupation for co-operation. The Israeli–PLO accords must be judged, rather, with an eye to the substantive material changes wrought by them. Evaluated thus, the evidence suggests that the Oslo II water accords did not really 'step' anywhere; this 'transitional arrangement' was a transition in little more than name.

Given all this, it should come as no surprise that Israeli–Palestinian 'co-operation' over water issues has continued since the breakdown of the peace process. During the first few months of the intifada there were no meetings of the JWC.[95] In January 2001, however, the JWC made a joint declaration urging people to keep water infrastructures 'out of the cycle of violence'.[96] Since then the JWC has met, albeit irregularly, and has been discussing and approving new projects. Moreover, the PWA is still approving new supply lines to Israeli settlements: during early 2002, for instance, approval was granted for an 11 km and 32-inch pipeline from the Green Line to Gush Etzion.[97] For practical reasons, the JSETs system has not been functioning at all.[98] However, for the most part the conventional breakthrough to breakdown narrative simply does not apply to Israeli–Palestinian water relations.

It should not be thought, however, that the discursive changes brought by Oslo simply sit atop material realities without impacting on them, or that the dressing up of Israel's domination of the West Bank's water resources, systems and supplies in liberal and legalistic terms had no significant material or institutional effects. To the contrary, the recent material improvements in water supplies to many of the West Bank's Palestinian communities have been largely attendant upon the discursive dressing up of occupation as 'co-operation'. The Oslo agreements and process as a whole bestowed a newfound legitimacy on Israeli–Palestinian relations (or at least signalled a gradual move towards legitimisation of these relations, the 'peace process'). Persuaded of the existence of the 'peace process' as a whole, and in 1995 of a 'breakthrough' in Israeli–Palestinian water relations, international donors were suddenly willing to take over from Israel the burden of ameliorating the critical water situation in the Palestinian territories. International donor moneys, in turn, have directly resulted in improved water supplies to many West Bank Palestinian communities. These improvements arguably followed less from the precise terms of the agreement – many of which are of illusory significance, and others of which are subject to an Israeli veto and to Israel's military power – than indirectly from the agreement's legitimacy in the eyes of the international community.

Much of what the Oslo II water accords directly achieved was discursive, insubstantial and altogether illusory. To speak of Israeli–Palestinian 'co-operation' in the water sector is to use no less than a misnomer. This is not, however, simply because 'the outcome of co-operation between an elephant and a fly is not hard to predict,' as Chomsky so pithily writes (since this is to assume that 'co-operation' represents a valid descriptor for Israeli–Palestinian relations), but because under Oslo, 'co-operation' has often been only minimally different from the occupation and domination that went before it.[99] Co-operation, in this context, is above all an internationally pleasing and acceptable signifier which obscures rather than elucidates the nature of Israeli–Palestinian relations. Or, as Meron Benvenisti so succinctly remarks, '"cooperation" based on the current power relationship is little more than permanent Israeli domination in disguise'.[100] Israel's colonial and apartheid water policies continued throughout the Oslo period. The only real mystery in this is how this sorry situation could have come about – and it is to this important question that we now turn.

CHAPTER 5

Excursus: The Case of the Eastern Aquifer

It was noted in the previous chapter that the Eastern Aquifer was defined in the Oslo II Agreement as the only one of the West Bank's three underground bodies of water that was not yet being exploited to its full potential. Whereas the North-eastern and Western Aquifers of the West Bank were being fully exploited to the tune of 145 and 362 mcmy respectively, the Eastern Aquifer alone had further potential for development:

> Schedule 10: Data Concerning Aquifers
>
> The existing extractions, utilization and estimated potential of the Eastern, North-eastern and Western Aquifers are as follows:
>
> Eastern Aquifer:
> * In the Jordan Valley, 40 mcm to Israeli users, from wells;
> * 24 mcm to Palestinians, from wells;
> * 30 mcm to Palestinians, from springs;
> * 78 mcm remaining quantities to be developed from the Eastern Aquifer
> * Total = 172 mcm
>
> All figures are annual average estimates.
> The total annual recharge is 679 mcm.[1]

These figures would come to have urgent planning and policy-making implications. The Oslo II Agreement stipulated that all of the Palestinians' immediate and future water needs (estimated at 70 to 80 mcmy) would be met through development of 'the Eastern Aquifer and other agreed sources in the West Bank,' it being largely on the strength of these figures that this stipulation became possible.[2] Moreover, Oslo II stated that this data would now 'constitute the basis and guidelines for the operation and decisions of the JWC,' the Joint Water Committee.[3] Yet the problem, as also briefly noted in the last chapter, is that there is strong evidence to suggest that these figures vastly overstate both the total yield and the remaining potential of the Eastern Aquifer. How could this have happened, and what are its implications?

This chapter tries to answer these questions by investigating in some depth the case of the Eastern Aquifer. Doing so involves deviating to some degree from the strongly politically oriented accounts of the previous and subsequent chapters, and also necessarily involves engaging with some rather technical and scientific issues. However, we make this brief excursus for three empirical reasons: in order to consider in more detail whether the Oslo II Agreement did indeed overstate the yield of the Eastern Aquifer; to examine how it was that these exaggerated values came to be enshrined within the Oslo II Agreement; and to assess the likely repercussions of these over-the-top figures both for the task of ameliorating the Palestinian water crisis in the southern West Bank, and for the environmental state of the Eastern Aquifer. In the process of considering these questions, this chapter looks back to the previous one, as well as forward to the analyses of the Oslo negotiations and water development work provided in chapters 6 and 7 respectively. The chapter also serves to emphasise the ineluctably techno-political character of hydrological knowledge. Such knowledge, I argue, is always endemically uncertain, and its production is always a social, political and institutional as well a technical matter.

I begin simply by describing how the Eastern Aquifer yield value given in the Oslo II Agreement was arrived at, and by detailing the major assumptions on which this value was based. I then raise some questions regarding the extent, and indeed the very existence of the Eastern Aquifer. And I conclude by considering some of the empirical and theoretical implications of the foregoing analysis.

SIX ASSUMPTIONS

Prior to the Oslo II Agreement, there had been a considerable measure of scientific disagreement as to the recharge and safe yield of the Eastern Aquifer. Where Oslo II put the Aquifer's recharge at 172 mcmy, Joshua Schwarz, a senior water manager at Israel's then state-owned water planning company Tahal had, just a few years earlier, estimated its yield at 100 mcmy, and before that at an uncertain 85–125 mcmy.[4] Most experts and commentators writing during the early and mid-1990s adopted figures somewhere in between this low value of 100 mcmy, and the much higher value that was later endorsed by Oslo II, with many of them choosing a figure of around 120 mcmy.[5] How then was it that the figure of 172 mcmy came to be officially validated within the Oslo II Agreement?

A simple answer would be that it was arrived at by Yossi Guttman and Ze'ev Golani of Tahal, the very same Israeli planning company which only a few years earlier had given the Eastern Aquifer's yield as 100 mcmy.[6] Guttman and Golani submitted their figures to Israel's Oslo II water negotiators, who in turn annexed them to the Oslo II Agreement. I will say more in the next chapter about why these Israeli-derived figures were not rejected or modified during the course of negotiations with the Palestinians. What concerns me here, though, is that these figures reflected a series of scientific, technical, social and political choices and assumptions. Six such choices and assumptions need detailing.

To start with, the figure of 172 mcmy was the product of a particular methodology. There are, at simplest, two ways of estimating the safe yield of a groundwater resource, either by measuring recharge or by measuring ouputs. To clarify, 'recharge' is that volume of water which, having fallen as precipitation, then filters down into underground aquifers, while the 'annual recharge' of an aquifer is thus that volume of water which falls as precipitation over it, minus that which returns directly to the atmosphere courtesy of evaporation and transpiration, and minus also that which simply flows over the land surface or through the soil. Following a recharge methodology, safe yield is calculated as precipitation minus evapo-transpiration minus surface flow. Following an output methodology, by contrast, yield is calculated as the sum total of spring discharge and well extraction levels. The Oslo II figure for the safe yield of the Eastern Aquifer was arrived at through the second of these methodologies, by totalling the yearly volume of spring discharges and well extractions across the Eastern Aquifer.[7]

Secondly, one problem with such a methodology (though with a recharge method even more so) is that there are tremendous practical difficulties and uncertainties involved in effecting it. Spring discharges in particular are incredibly awkward to keep track of. The problem is not only that monitoring springs is so time-consuming, or that many countless smaller ones go unmonitored; the problems also arise from the fact that springs are often technically challenging to monitor, and are highly changeable in the level of their discharges. In the case of the Eastern Aquifer, most of the 78 mcmy of remaining water mentioned in the Oslo II Agreement flowed from a series of springs located along the shore of the Dead Sea. These Dead Sea springs – in particular the largest of them, Ayn Fashkha – take the form of a mass of tiny seeps and rivulets, the locations of which are constantly changing in accordance with discharge levels. Discharge levels themselves vary enormously from one year to the next. In consequence, Ayn Fashkha is almost impossible to measure with any accuracy or consistency, so much so that during recent years only Tahal and the IHS have even attempted to do so.[8] When Tahal measured Ayn Fashkha in the late 1980s, a flow of 40 mcmy was recorded. By contrast, when the IHS repeated the exercise in 1992, they recorded a flow of 80 mcmy, double the earlier figure.[9] The value given in Oslo II for the Eastern Aquifer assumed a combined discharge from all of the Dead Sea springs (Ayn Fashkha plus the others) of 80 mcmy – this figure being closer but not equivalent to that produced by the IHS.[10] If Guttman and Golani had followed the IHS's figures to the letter, they could feasibly have arrived at a safe yield for the Eastern Aquifer of well over 180 mcmy; if they had followed Tahal's earlier work, their figure could have been as low as 140 mcmy. There is significant uncertainty here. Many knowledgeable experts are of the opinion, however, that the IHS's figures were arrived at on the back of previous heavy rainfall, and that Tahal's far lower figure is probably a more accurate representation of annual average discharge levels from the Dead Sea springs.[11]

Thirdly, the output approach to measuring safe yield inevitably involves assumptions about the stability or otherwise of the underground water table, since yield is only equivalent to total outputs if underground water levels remain constant. In the case of the Eastern Aquifer, the yield value was arrived at simply by totalling spring flow and well extraction levels, with no adjustment being made to allow for variations in the water table level.[12] However, as noted in the previous chapter, there is strong evidence that water table levels in the Eastern Aquifer have been dropping sharply during recent years, with water levels in one of the wells, Herodian 3, just south of Bethlehem, having been declining at a rate of over 5m per year since 1981.[13] According to the IHS, the water table of the 'Herodian Aquifer' (as they call the Herodian area of the Eastern Aquifer) dropped by 1.75m between 1972 and 1996.[14] This does not mean that the water table is dropping throughout the Eastern Aquifer; what it does suggest, though, is that the aquifer's discharge and extraction levels are probably very different from its recharge.

Fourthly, the figure of 172mcmy is premised on assumptions about the technological and economic feasibility of exploiting new and additional waters from the Eastern Aquifer. Most of the aquifer's as yet undeveloped waters are discharged, as noted above, from springs located alongside the Dead Sea. The problem, however, is that these spring waters are highly saline. The Oslo II figures were premised on the assumptions that these waters flow down from sweet rock strata well above the Dead Sea, that they only become saline once they reach the floor of the Jordan Valley, and furthermore that it is possible to intercept these waters before they become salinised.[15] It appears to be the case, however, that some of the water discharged along the shore of the Dead Sea comes from much deeper aquifers, where it is already saline, and indeed that some of this water 'fell as rain 25,000 years ago, possibly as far away as the Atlas Mountains'.[16] Even those waters that do flow down from sweet rock strata above the Dead Sea may prove impossible to exploit. One recently developed model of the Eastern Aquifer concludes, for instance, that the salt water that emerges from the saline Dead Sea springs journeys there, first of all, by slowly filtering into the Jordan Valley, and thereafter by flowing southwards alongside the Jordan River or northwards along the Dead Sea to the points where it is discharged.[17] The implications of this are that in the areas where the water is still sweet – that is, above the floor of the Jordan Valley – it may flow so thinly as to render exploitation economically, and perhaps even technologically, unfeasible.[18] Whether this is the case or not, only time will tell. Yet what is clear is that the Oslo II figures for the safe yield of the Eastern Aquifer, and for the quantities to be developed from it, both involved the assumption that these additional waters could feasibly be exploited. If it had not been judged possible to exploit the Dead Sea spring waters, then the yield value for the Eastern Aquifer would have been a great deal lower.

A fifth and related issue regards the feasibility of exploiting these additional waters without incurring environmental damage to the aquifer itself. The danger here is that if these erstwhile spring waters were fully exploited, salt water from the Dead Sea and the floor of the Jordan Valley might well flow up into the lower stretches of the aquifer, and perhaps even contaminate wells pumping from it, thereby rendering these wells saline and unusable.[19] In light of this, Yossi Guttman suggests that at least 20mcmy should always be allowed to flow from the Dead Sea springs, and thus discounts the possibility that 78mcmy could ever be exploited from the Eastern Aquifer on a sustainable basis – this coming, remarkably, from one of the two Israeli hydrologists who produced the figure of 78 mcmy for the Israeli Oslo II negotiators.[20]

Sixth and lastly, the figures given in the Oslo II Agreement reflect a set of economic, social and political priorities. This, indeed, is intrinsic to the very concept of 'safe yield'. While in most commentary on the Middle East's water problems, 'safe yield' is 'usually considered to be equal to the annual recharge rate,' the truth is rather that the concept is contested and inevitably value-laden.[21] For instance, Meinzer classically defined safe yield as 'the rate at which water can be withdrawn from an aquifer for human use without depleting the system to such an extent that withdrawal at this rate is no longer economically feasible'.[22] Todd's alternative and somewhat later definition, while moving away from this strictly economic emphasis, nevertheless revolved around questions of value: as he put it, the 'safe yield of a ground water aquifer is the amount of water that can be withdrawn from it annually without producing an undesired result'.[23] The very notion of 'safe yield', then, involves assumptions about economic and social 'desirability' (it is, as Todd puts it, 'essentially meaningless from a hydrological standpoint' – an administratively useful problem-solving concept that 'in spite of the reservations of many hydro-geologists ... must be applied whenever the use of an aquifer is planned or managed').[24] Given this, it is inevitable that particular safe yield values will likewise be premised upon various value assumptions. In the case of the Eastern Aquifer, safe yield values for it are premised on certain assumptions about the environmental value of the Dead Sea and its springs. The Dead Sea springs are important nature reserve areas, ones that the Israeli environmental lobby, in particular, is very keen to preserve. According to Joshua Schwarz, his earlier yield value of 100 mcmy was premised on a judgement about the influence and strength of Israel's environmental lobby, while the very different value given in the Oslo II Agreement reflected a quite contrary assessment of the environmental acceptability of seeing these springs shrink and disappear.[25] More broadly, all safe yield values for the Eastern Aquifer implicitly assume that the retreat of the Dead Sea is not undesirable. The level of the Dead Sea is currently declining at a rate of 0.8 m per year, and while this is mainly as a result of water being diverted from the upper Jordan River, to a lesser extent it also follows from exploitation of local aquifers.[26] Increased exploitation of the Dead Sea springs would inevitably speed up the decline of the Dead Sea itself. Yet the Oslo II figures, and all other expert discourse besides, takes no account of the Dead Sea's plight – with consideration of the Dead Sea being routinely absent from donor-funded environmental assessments, for instance.[27] It should almost go without saying that this is because the Dead Sea is a highly saline body of water, one that is situated over 400m below sea level, and that has negligible economic value as a source of useable water. The Oslo II figures, in sum, reflect a mesh of social assumptions both regarding the instrumental value of the Dead Sea, and regarding the environmental-political importance of the saline springs that lie alongside it.

CONSTRUCTING BOUNDARIES

If all this uncertainty is disconcerting, it becomes all the more so when one learns that the Eastern Aquifer is in some sense a social construct. The Eastern Aquifer, we will recall, is usually defined as one of three such bodies of groundwater underlying the West Bank (see Figure 1.3). Less often noted, however, is the fact that there is a great deal of scientific uncertainty regarding its precise spatial extent (compare, for instance, the four maps shown together in Figure 5.1). Just as there is uncertainty over the safe yield of the Eastern Aquifer, so too is there uncertainty over its boundaries.

The reasons for this are several. To start with, the three West Bank aquifers are usually thought of as being delineated in terms of the direction of groundwater flow (and indeed Figure 1.3 and most other maps convey precisely this impression). The problems in delineating the Eastern Aquifer's boundaries in this way are several, however. First, because of their karst geology, and owing to the tectonic influences of the Jordan Rift Valley, the West Bank's underground systems are laced with a complex array of anticlines, synclines and faults, such that water flowing within the Eastern Aquifer moves not in a uniformly eastward direction, but in the chaotic manner of Brownian molecules. Although the water within the Eastern Aquifer does generally flow in an eastward direction, it does not do so in any consistent or linear fashion, and thus attempts to demarcate the aquifer's borders inevitably involve a degree of simplification. Second, the general direction of groundwater flow can vary with depth, and especially between one water-bearing layer and the next, such that divides (if based upon general direction of groundwater flow) cannot be accurately represented in two-dimensional form. Third, the general direction of flow can vary over time as a result of changes in rates of precipitation and infiltration, and also as a result of human activity – to the extent that heavy pumping in the Eastern Aquifer would result in its boundary with the Western Aquifer drifting

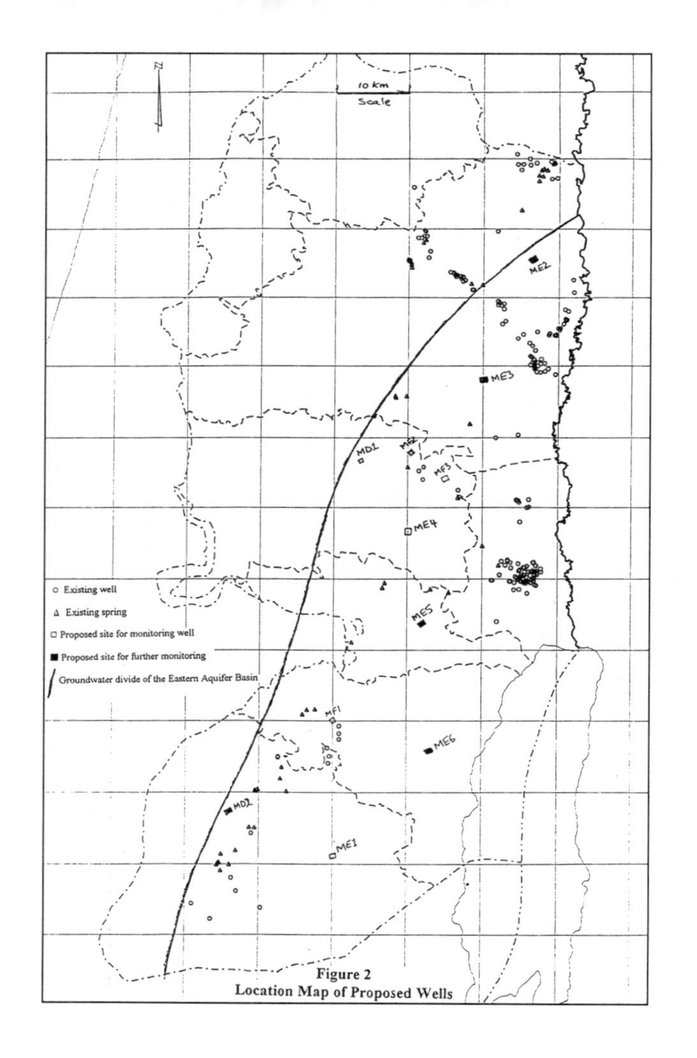

Figure 2
Location Map of Proposed Wells

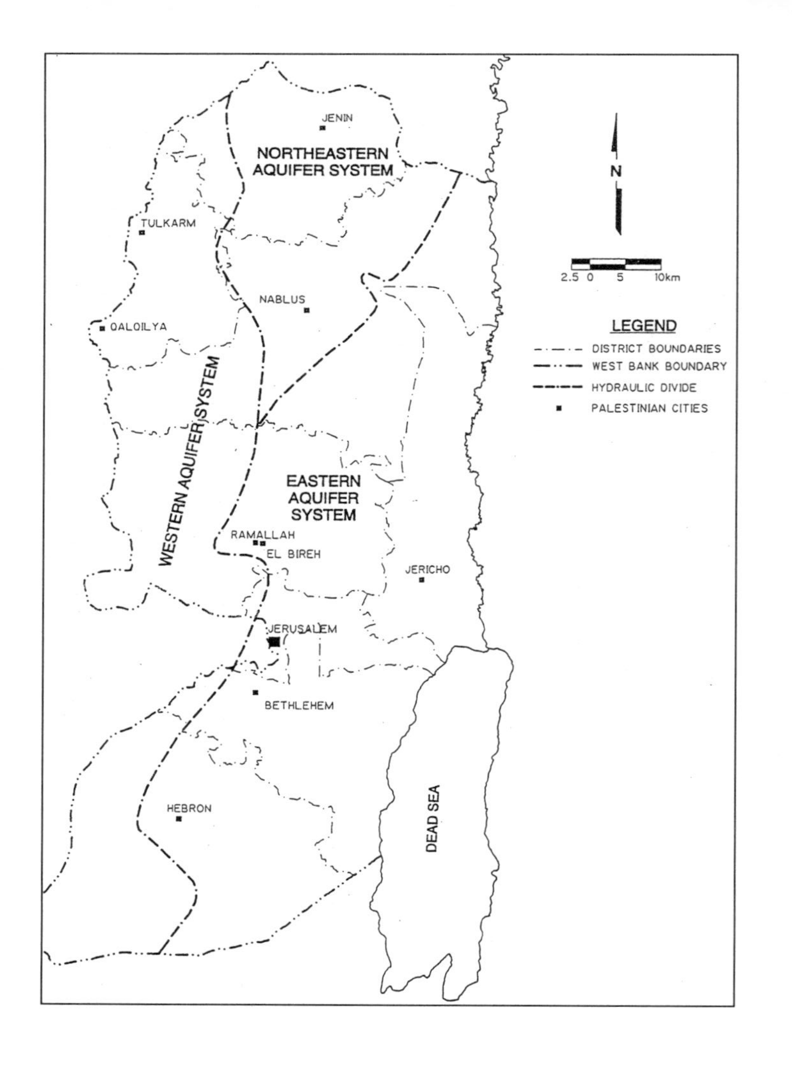

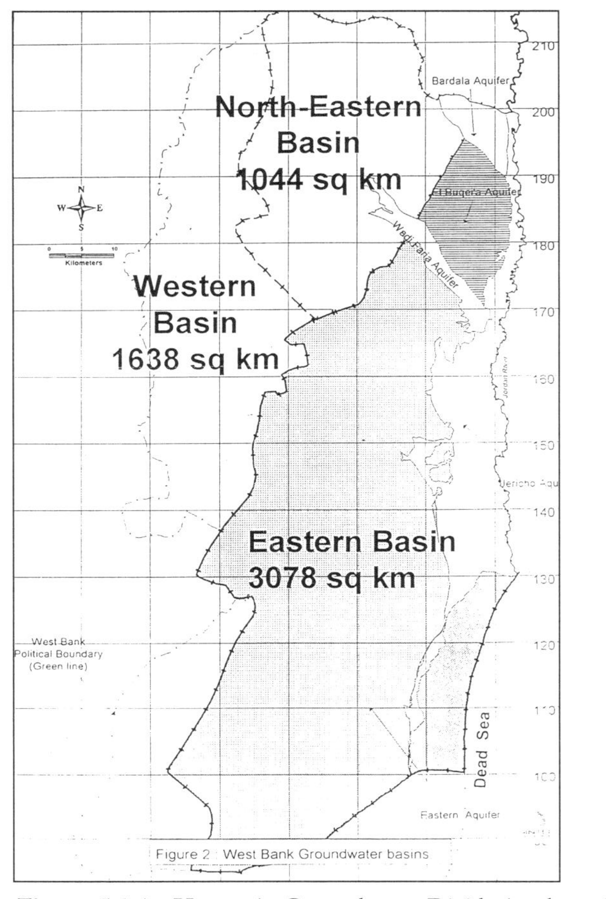

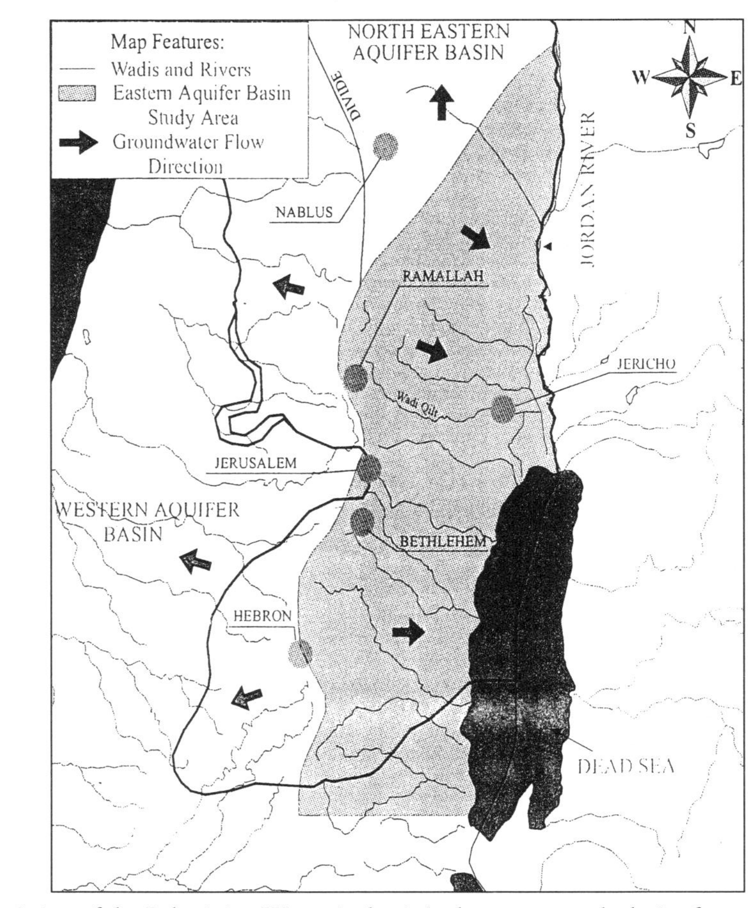

Figure 5.1 An Uncertain Groundwater Divide (made available by permission of the Palestinian Water Authority); the maps are, clockwise from the top left: CDM/Morganti, Task 20; CDM/Morganti, Task 19; CDM/Morganti, Task 18; ANTEA, Well Development Study

eastwards.[28] For each of these reasons, there are no absolute or unchanging limits to the Eastern Aquifer, and attempts to represent it inevitably involve a large measure of simplification.

Beyond this problem of representing complex realities in simple cartographic form, an even more significant problem lies in the inaccessibility of hydro-geological structures and processes. It is well-nigh impossible to delineate boundaries of the general direction of groundwater flow – even at a given moment in time, and even within a given water-bearing layer – simply because hydro-geologists lack much of the data necessary for representing these hidden realities accurately. With these difficulties in mind, groundwater divides are in practice often delineated not in terms of the direction of groundwater flow, but in accordance with either surface flow, or structural geology. It so happens that the West Bank's three aquifers were originally delineated in terms of regional structural geology, for the simple and contingent reason that the British engineering consultants who first studied the West Bank's hydrology – on a project for the Jordanian government – had a structural geologist among their survey team.[29] If the West Bank's groundwater systems had been demarcated by a different group, and using a different methodology, these systems might well not have been defined as consisting of three discrete aquifers. To this extent, the tripartite division of the West Bank is a historically contingent social and applied scientific construct. The very existence of a distinct natural object called the 'Eastern Aquifer' can quite reasonably be called into question.

IMPLICATIONS

It cannot be stated with absolute certainty that the figure of 172 mcmy given in the Oslo II Agreement overstates the safe yield of the Eastern Aquifer. There can be no absolute certainty about such matters: new subterranean spring discharges may be discovered, new deep fault lines may be tapped, and technological developments may render novel types of extraction possible. However, in light of the evidence reviewed above, it seems evident that, at present levels of technological development, much less than the stated remaining 78 mcmy could feasibly be exploited from the Eastern Aquifer – without, that is, doing it irreparable ecological damage. If we are to follow Guttman's contention that a minimum of 20 mcmy must be allowed to continue to flow from the Dead Sea springs in order to prevent the aquifer's salinisation, then the remaining water available for development drops to below 60 mcmy. If we are also to adopt a lower figure for the discharge of the Dead Sea springs, as many experts recommend, then the remaining water available becomes perhaps less than 40 mcmy. If we take into account the possibilities that much of the water discharged from the Dead Sea springs may derive from inaccessibly deep and saline aquifers, or may flow in a manner that makes it impossible to exploit before becoming saline, then the yield value could drop further still. If we bear in mind that the Eastern Aquifer is an uncertain social construct, one that perhaps consists of relatively autonomous sub-aquifers, then we may well conclude that certain parts of the aquifer – and the Herodian wellfield in particular – are already being exploited at unsustainable levels. It seems reasonably clear that in the Oslo II Agreement the Palestinians were promised a new source of water that simply was not fully there for the taking.

How then did these uncertain but seemingly exaggerated figures for the Eastern Aquifer come to feature in the Oslo II Agreement? In part, it must be said, this was because of the amount of complex work and the variety of methodological (and also social) assumptions that necessarily go into calculating safe yield values. Discussions of data problems in relation to Middle Eastern water issues typically operate with an 'information deficit' level of scientific knowledge, with the main problems being perceived as arising, for instance, from 'the use of data by parties with their own vested interests ... sloppy reporting by analysts, researchers, government agencies and the media,' as well as from a general 'lack of data'.[30] But the uncertainties considered here could never be wholly resolved through more accurate monitoring and modelling, since they are the products of often arbitrary choices and assumptions, as well as of value commitments. In part, then, the reason why the Oslo II Eastern Aquifer figures are so dubious and contested is simply because the uncertainties are so endemic.

Beyond this, however, these figures were borne either out of misunderstanding and miscommunication between Yossi Guttman, Ze'ev Golani and the Israeli negotiating team, or because of outright manipulation on the part of the latter. I leave it to the reader to judge which of these is the more likely. 'They take our numbers, and they do what they understood at the time,' observes Yossi Guttman of the Israeli negotiators.[31] 'The Israelis fabricated the facts about aquifer yields ... to serve their negotiating position,' observe two Palestinian water experts (who also claim that Israel manipulated the figures for the Western Aquifer in order to be able to claim that it was already being fully exploited).[32] Certainly it is a strange coincidence that a time when the Israeli state was searching for new water supplies for the Palestinians, and for a way of denying the Palestinians rights to develop the Western and North-eastern aquifers, that it suddenly managed to conjure up a sparkling new and until then barely noticed water resource.[33] Such are the wonders of modern state power.

Since 1995, the Oslo II data has become a constant reference point for commentaries, analyses, plans and proposals on Israeli–Palestinian water issues.[34] Enshrined within a politically important document, and validated by the two parties, the figure of 172 mcmy has become the officially recognised yield for the Eastern Aquifer. While many Palestinian experts in particular continue to question its validity, it has for most purposes become 'black-boxed' as a technical truth without a history.[35] The same can be said of the Eastern Aquifer itself which, despite being a social construct has now become an accepted natural object, to potentially crucial effect.

Since 1995, and on the strength of the data contained in the Oslo II Agreement, numerous deep wells have been drilled into the Eastern

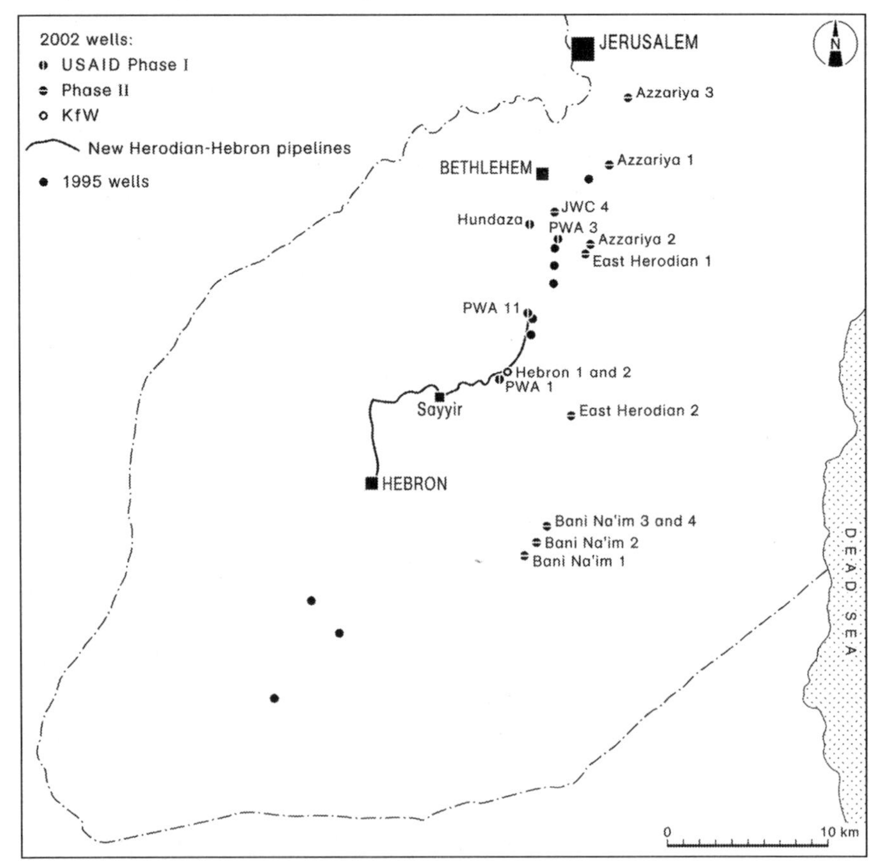

Figure 5.2 New Palestinian Wells in the Southern West Bank

Aquifer, at least six of them so far into the Herodian wellfield (see Figure 5.2). Still other wells are planned. However, as a study conducted in 1998 for the PWA concluded, if all of the then planned wells were to be drilled and brought into operation, there would be an estimated decline in the Herodian wellfield water table of up to 120 m over a four-year period.[36] Existing Eastern Aquifer supply wells would possibly dry up. There could even be wider impacts on the Western Aquifer, with a drop in the Eastern Aquifer water table potentially leading to leakage from the Western Aquifer, and in turn to a lowering of the Western Aquifer water table.[37] In light of these dangers, two Palestinian water experts argue that there 'is a worry that drilling a large number of production wells in the Herodian Wellfield will cause severe negative impacts on the productivity of the Wellfield as a whole and the consequences might be disastrous'.[38] A leading Palestinian NGO contends 'that the donor-driven funding rush guided by the aim to support the peace agreement has encouraged potentially unsustainable development of regional water resources,' and 'urges the Palestine Water Authority to refrain from any further production well drilling'.[39] There is a great deal of uncertainty here. It is quite possible, though, that Israel's unwillingness to forego all but a smattering of its current water supplies, combined with the PWA's desperation for new wells and international donors' continuing desire to pump money into the peace process, might lead to the salinisation and destruction of the Eastern Aquifer. Such an outcome may not be many years away. And if that were to happen, the long-term consequences for the Palestinian population of the West Bank and Gaza would be stark indeed.

CHAPTER 6

Explaining Oslo

The initial Oslo breakthrough and the negotiations that led to it have often been explained in strongly personalistic terms, as the products of a special chemistry – the 'Oslo spirit' – that supposedly developed amidst the woods and fjords of Norway.[1] Hopefully, though, it should now be clear that, regardless of whether any such benevolent spirit actually existed, the Oslo Accords and process need to be explained in thoroughly structural terms, with an eye to the long-term projects, strategies, policies and powers of the Israel state and the PLO. Some of these structural contexts have already been explicated: we have seen, for instance, that the specific character of the colonial encounter between the Zionist movement and the Palestinians led to the development of an Israeli state and society that placed great emphasis on land and water, and that thus sought to deprive the Palestinians of land and water to as great an extent as possible; and we have also seen that, after 1967, this colonial project was extended and radicalised, leading to the development of an institutionalised apartheid system that, amongst other things, ensured that Israeli settlers received constant supplies of water, while Palestinian communities often went weeks, even months, without. These and other parallel historical contexts are of central importance in understanding and explaining developments under Oslo. Nonetheless, we still need to consider in general terms why the Oslo process was initiated, and why it ended up taking the broad form that it did; and we need to ascertain, more particularly, what the motivations of Israeli and Palestinian water negotiators were in constructing a West Bank water regime that differed so minimally from that which had preceded it. An instant reaction to this last question might be that Israeli ambitions have been all too clear. But what of the Palestinian motivations? Why, we may ask, was the PLO willing to agree to terms that enabled Israel to dress up domination as co-operation? Why were the Palestinians unable to reject or modify Israel's dubious hydrological statistics? It is to such questions that the present chapter turns.

As previously, I begin at a general level, considering the various aims, ambitions and power relations that informed the Oslo agreements and process as a whole. Only in the second section do we turn to the more specific issue of water.

THE ISRAEL–ARAFAT AGREEMENT

Ever since occupying the West Bank and Gaza in 1967, Israeli elites have been well aware of their looming 'demographic problem' – that of how to avoid absorbing too many Palestinians into the Jewish state and thereby undermining its definitively Jewish character. Arnon Soffer, speaking in 1988, voiced precisely this concern:

> The number of Palestinian Arabs in the territories and in Israel now reaches 2.2 million, while the number of Jews is 3.5 million. In 12 years, the Arabs will reach 3.5 million while the Jews will reach 4.2 million. It does not matter whether the Arabs will be 44 or 46 percent of the population. What matters is that it will be a binational state. Whoever brings about this situation will be responsible for the end of the Jewish, Zionist state.[2]

Under occupation, the main response to this problem was to keep the West Bank and Gaza Strip under military rule, thereby denying the newly occupied Palestinian population Israeli citizenship, and preserving the demographic viability of the Jewish state. But this uneasy situation could not be maintained forever. One suggested alternative was some form of limited territorial compromise. The Allon Plan, for instance, envisaged Israeli annexation of a third or more of the West Bank, with the remainder, and most importantly the heavily Palestinian-populated areas, being returned to Jordan.[3] Later plans advocated limited Palestinian autonomy within the West Bank and Gaza. The 1978 Camp David Accords, for example, called for Palestinian autonomy in terms which were very similar to those of the Declaration of Principles.[4] More significantly still, the Baker-Shamir-Peres Plan of 1989 called for 'free and democratic elections' for the Palestinians living in the Territories, leading to Palestinian 'autonomy' over 'affairs of daily life', with Israel retaining control of security, foreign affairs and all matters relating to settlers.[5] This plan noted that there would be no 'additional Palestinian state in the Gaza district and in the area between Israel and Jordan,' the supposition being, as Chomsky observes, that the Palestinians already have their own state: Jordan.[6] This plan was approved by the Israeli Knesset, and became both the formal basis of Israeli policy, and a formative influence on the Oslo process. Israel, in short, had long been considering ways of restructuring its control over the West Bank and Gaza, and of holding elections therein that would afford its occupation and colonisation a degree of legitimacy.

Besides this long-term demographic problem, Israeli business elites had also started campaigning for a peace agreement of some sort with the Palestinians. Since the 1970s, Israel had been undergoing tremendous social, political and economic changes, with the fracturing of the state-centred Labor Zionist consensus that had dominated Israel since its inception, and the genesis of new civil society movements of both religious-nationalist and liberal variants.[7] On the one hand these transformations had seen the rise of Gush Enumin and other religious and ethnically defined political movements and parties. On the other, though, they had also involved the partial liberalisation of the Israeli economy away from the etatist institutional forms inherited from the pre-state days (the decline of the Histadrut trade union, for instance), and concomitantly, the emergence of liberalising elites who no longer defined their (or Israel's) interests in traditional national security terms.[8] For the first time in the late 1980s, these liberal business leaders started entering into public political discourse, arguing that the Israeli economy needed to attract foreign capital and find new (especially Arab) markets, and that moves towards peace were an essential precondition for this. Many of them also argued, unsurprisingly, for peace terms that would ensure continued economic hegemony over the Territories: as the President of the Israeli Manufacturers' Association observed, 'it's not important whether there will be a Palestinian state, autonomy, or a Palestinian–Jordanian state. The economic borders between Israel and the territories must remain open.'[9] By no small coincidence, many of the key Israeli architects of the initial Oslo agreement (Shimon Peres, Yossi Beilin and Yair Hirschfeld, for example) were also leading proponents of economic liberalisation; the secret Oslo negotiations, moreover, were conducted entirely by civilians, and without the prior knowledge of the IDF.[10] It was for these reasons that the agreement came to place such emphasis on economic co-operation (one commentator even goes so far as to suggest that the Declaration of Principles was 'primarily an economic document').[11] The initial Oslo agreement and its terms were at least in part the result of the growing liberalisation of the Israeli political economy.

A third and more immediate reason for the Oslo Agreement was Israel's growing security problems in the Territories. The onset of the intifada in 1987 had rendered Gaza and some of the West Bank cities practically ungovernable; moreover during the early 1990s, guns had started replacing stones as the most prominent means of Palestinian resistance, and the security of Israeli settlers and civilians had become increasingly under threat. Influential Israeli figures such as Ze'ev Schiff thus started to advocate 'security for peace' plans for the Territories, with the Palestinians being granted functional autonomy with their own 'large police force', linked in confederation with Jordan.[12] Yitzhak Rabin, very much a traditional Labor Zionist 'security hawk', not an economic liberaliser, became persuaded that Israel's security interests could best be met through indirect control, rather than direct occupation. Gaza, no matter how much he wished it, was not about to sink into the sea. He thus set about negotiating a deal with the PLO that met Israel's security requirements (this process happening in tandem with the closure of the West Bank and Gaza, as well as heightened levels of military repression), but that did not contradict Israel's territorial and economic objectives.[13] This is precisely what the early Oslo agreements achieved. Rabin himself was candid about the security logic that informed the agreement, noting that the 'Palestinians will be better at it [enforcing order] than we were because they will allow no appeals to the Supreme Court and will prevent the Israeli Association of Civil Rights from criticising the conditions there by denying it access to the area. They will rule by their own methods, freeing, and this is most important, the Israeli army soldiers from having to do what they will do.'[14] While the economic liberalisers were undoubtedly influential in pushing for negotiations and in moulding the initial Oslo agreement, it was this security logic of Rabin and the IDF that ended up being the more influential. The liberalisers just wanted peace; the military establishment was to define its terms. Indicatively, while Peres and Beilin were the central figures during the secret Oslo negotiations, the later talks that led to the agreement of the far more detailed Gaza–Jericho and Oslo II Accords were led primarily by Rabin and the IDF. The result was the creation of a client authority whose functions were primarily repressive, which developed an appalling human rights record, and which came to have the highest proportion of police to civilians of any political authority in the world.

It is routinely assumed that the 1992 Israeli elections – and the subsequent formation of an Israeli government that for the first time in 15 years was not led by the Likud – were key to the negotiation of the Oslo Agreement. Certainly it is true that Likud leaders have generally been extremely loath to make concessions on the West Bank and Gaza. Begin never had any intention of implementing the Camp David agreement on Palestinian autonomy, and as for Yitzhak Shamir (who preceded Rabin as Israeli Prime Minister), his aim during the pre-Oslo Madrid negotiations was, as he acknowledged on leaving office in June 1992, 'to drag out talks on Palestinian self-rule for ten years while attempting to settle hundreds of thousands of Jews in the Occupied Territories'.[15] Yet while it is unthinkable that the original Oslo Accord could have been negotiated and agreed by a Likud government, it would be a mistake to overstate the differences between the centre left and centre right of the Israeli political spectrum. Labor and Likud governments since 1977 have equally supported and extended settlement-building programmes. They have both put continual pressure on the PA to extend its repression of Palestinian opposition groups. They have equally favoured the maintenance of economic hegemony over the Territories. Both have been willing to transfer pockets of territory to the PA within the limited framework set out by Oslo – and if the negotiations under Binyamin Netanyahu's Likud administration were much more fraught than those that preceded them, this was in large part because Israel had already redeployed from most Palestinian population centres, making the task of further redeployment much more complex (Ehud Barak found this problem equally taxing). Indeed, the most striking differences between Likud and Labor governments in their relations with the Palestinians have generally lain not in substantive issues of policy, but in presentation and tactics, with Likud appealing to a right-wing constituency and thus being prepared to make openly expansionist and nationalist gestures, and Labor adopting generally dovish rhetoric, yet undertaking policies that are usually very similar.[16] The ebb and flow of the Oslo process, in short, has not simply corresponded to whether Labor or the Likud has been in power. To the contrary, this process needs to be understood above all as the result of established territorial and security strategies, over which there has been a relatively high degree of national consensus.

In agreeing to conform to this consensus, Yasser Arafat and the PLO deferred their own stated policy, calling for the establishment of a Palestinian state on the whole of the West Bank, including East Jerusalem, and Gaza Strip. They also departed from the consensus position of the international community – which included in this case European and most Arab states, and Russia and the former USSR, but which excluded the US and Israel – calling explicitly for a two-state solution to the Israeli–Palestinian conflict. The PLO had formally become party to this consensus position in 1988, when it had accepted UN Security Council Resolution 242 (calling for 'withdrawal of Israel from territories occupied' in 1967), had stated a willingness to recognise Israel, and had declared a nominal independence over the West Bank, including East Jerusalem, and Gaza Strip (prior to that the PLO had refused to accept 242 on the grounds that it made no reference to Palestinian national rights).[17] The Declaration of Principles marked a retreat from this 'accommodationist' position, in favour of Israel's 'rejectionist' denial of Palestinian rights to self-determination.[18] The agreement did admittedly define Resolution 242 as the formal basis for negotiations, but this did not imply recognition of Palestinian rights, nor did it imply for Israel or the US that withdrawal would take place to pre-1967 borders (Israel and the US argue that 242's reference to withdrawal 'from territories occupied' – as against withdrawal from 'the territories' – implies that the resolution does not call for full withdrawal to the pre-1967 borders).[19] Moreover, the agreement neither mentioned Palestinian statehood, nor recognised Palestinian national rights. Rabin and Peres agreed to the Oslo terms while being explicitly opposed to the idea of Palestinian statehood (as Rabin said, there is 'nothing [in the Accord] about a Palestinian state or a capital in part of Jerusalem. I stick to my position: no Palestinian state, Jerusalem must remain united under Israeli sovereignty and be our capital forever... I don't believe there is room for an additional state between Israel and Jordan').[20] Israel and the PLO exchanged letters of mutual recognition, yet while the PLO accepted 'the right of the State of Israel to exist in peace and security,' Israel conversely only recognised the PLO as 'the representative of the Palestinian people': Israel was recognised as a nation-state, the PLO as an organisation. Furthermore, the reference to 'the Palestinian people' was of no significance, since Begin had already accepted 'the legitimate rights of the Palestinian people' at Camp David, noting then that he interpreted this as referring not to a people with national rights, but to the inhabitants of the West Bank and Gaza (the US accepted Israel's interpretation).[21] The Declaration of Principles opened a process and an interim autonomy period, but involved no commitment on Israel's part to accepting Palestinian statehood.

The central principles of the Declaration of Principles were in many regards similar to those of the autonomy section of the Camp David

Accords. This likeness was indeed acknowledged by Israel's negotiators, who unveiled the agreement before cabinet and country as 'a major improvement over the Camp David Accords'.[22] Rabin had argued in the early 1980s that the PLO should be denied a role in negotiations 'even if it accepts all of the conditions of negotiations on the basis of the Camp David agreements, because the essence of the willingness to speak with the PLO is the willingness to speak about the establishment of a Palestinian state, which must be opposed'.[23] Yet by 1993 this argument no longer held – not, however, because Rabin had changed his views about Palestinian statehood, but because the PLO had 'moderated' its own position to conform to Israel's rejection of Palestinian rights. 'There has been a change in them, not us,' observed Shimon Peres, as the Oslo Agreement was announced; 'we are not negotiating with the PLO, but only with a shadow of its former self'.[24] The question that necessarily arises is why it was that Arafat and the PLO now accepted Israel's Camp David-style peace process.

Put bluntly, Arafat capitulated to Israel's rejectionist terms because he and the PLO were weak and increasingly desperate. They had always been dependent on diplomatic and especially financial support from Saudi Arabia, Kuwait and the small Gulf states, but following the 1990–91 Gulf War, when Arafat openly lent his support to Saddam Hussein, the greater part of this support was withdrawn.[25] Moreover, after the war the half million Palestinians who had been working in Kuwait prior to the Iraqi invasion were no longer welcome, and hence the majority of them left for Jordan; the PLO had long benefited from private remittances from Palestinians working in the Gulf, and these also started to dry up. The result was that the PLO found itself in financial crisis to the tune of $100 million a year, and started closing various of its diplomatic offices, including some of those at its headquarters in Tunis.[26] Financially and politically isolated, Arafat sought American sponsorship and a deal that would rehabilitate the PLO.

Quite aside from this, Arafat and the PLO in Tunis also faced internal Palestinian challenges from political movements and leaders within the Territories. The PLO had been based in Tunis ever since 1982, after Israel's invasion of Lebanon had forced it to leave West Beirut; housed in Tunis, the PLO was 1000 miles away from Palestine, and more ineffectual than ever. The intifada in part represented a response to this – an attempt by Palestinians within the West Bank and Gaza to take the lead in resisting the Israeli occupation – and thus constituted a challenge both to Israel and the PLO. This challenge came first of all from the Unified National Leadership of the Intifada, which was formed without Tunis's backing, and was only later controlled by it.[27] But it came increasingly too from emergent Palestinian Islamist movements, most notably Hamas, which as the intifada progressed took an ever more active role in resisting the occupation, and became a threat to the PLO's prestige, credibility and status as 'sole legitimate representative of the Palestinian people' (Hamas lay and still lies outside the PLO umbrella).[28] Thus arose a series of conflicts between the PLO 'outsiders' in Tunis, and Palestinian 'insiders' within the Territories.

With the onset of the Madrid peace talks in 1991, this inside-outside conflict took on yet another form. The PLO was denied official representation in these negotiations, and their place was taken instead by a PLO- and Israeli-approved delegation from the Territories (initially as part of a joint Israeli–Palestinian team). This delegation received international attention and acclaim as the media-friendly face of the Palestinians. Although under instruction from Tunis, they increasingly acted independently. Moreover, they held a firm line on many issues, being only willing to assent to a phased process, for instance, on condition that Israel accepted Palestinian self-determination and statehood as its final destination.[29] Shimon Peres judged that Arafat would be much more compliant than the inside negotiators, being 'convinced that if Arafat was allowed to return and rule in Gaza and Jericho... he would yield, for the time being, on virtually everything else. This included the Palestinians' core issues.'[30] Peres thus set about trying to establish secret contacts with Arafat. As for the PLO Chairman, the Oslo negotiations and process enabled him to bypass and undermine the Madrid delegation, and to re-establish control of the Palestinian national movement.

That Arafat was able to re-establish this control, and to push through the Oslo deal, testifies to how personalised his rule of the PLO had become. It had not always been like this. Autocratic as his leadership style had always to some extent been, PLO decision-making had nevertheless long been constrained by the demands of its Arab supporters. Moreover, important policy decisions were generally made, in accordance with the PLO's constitution, within the 400-member Palestine National Council (PNC) and the much smaller Palestine Central Council (PCC). However, after the Gulf War much of this changed. As the PLO's financial and political crisis deepened, so the influence of external backers decreased, and so the influence of Arafat and associates grew stronger.[31] Arafat was thus able to conduct secret talks without having consulted the PNC or PCC, and he signed the Declaration of Principles in Washington without even convening the PNC, an act that was evidently unconstitutional. It is with good reason that Chomsky refers to this Declaration as the 'Israel–Arafat Agreement'.[32]

Just as the Oslo Agreement was premised above all on Israel's security concerns and territorial ambitions, so too was it only arrived at because of the weakness of the PLO in relation to its unyielding and powerful adversary. This same power imbalance continued throughout the Oslo process to inform and structure Israeli–Palestinian relations. Once the PLO had accepted the initial Oslo terms, there was little that it could do to change this imbalance of power, or to redirect the process. It was up to Israel to decide what concessions it was willing to make – and if the Palestinians wanted the process to move along quickly, they would just have to accept them. Throughout the course of the negotiations, Israel's security doctrines thus always prevailed – as Uri Savir reports, Palestinian negotiators had received 'specific instructions from Arafat' to accommodate the Israelis on 'every aspect' relating to security.[33] Security, it might be noted, is seemingly the preserve of Israelis only.[34] The Palestinians were only granted rights to the extent that they complemented Israel's security, territorial and economic strategies, its perceived 'vital national interests'. Indeed, it may perhaps be said that the key site of political debate within the Oslo process lay not so much within Israeli–PLO negotiating forums, as within Israeli politics and society. Shimon Peres was especially candid on this matter, observing prior to the Oslo II talks that 'in some ways we are negotiating with ourselves'.[35] Reflecting a year later on the Oslo II Agreement itself, his candour had reached new heights: 'we screwed the Palestinians,' he noted, a remarkable admission coming from a Foreign Minister, and from someone so internationally renowned as one of Israel's leading doves.[36] It is within these contexts that the Oslo agreements and process need to be understood. These did not occur because of some outbreak of compassion or personal chemistry, and still less because the PLO had at last decided to recognise Israel (since this had happened long before), but because the weakness of Yasser Arafat and the Tunis-based PLO made them suitable candidates for appointment as Israel's legitimate quisling enforcers in the West Bank and Gaza.

NEGOTIATING WATER

With this broad picture established we can now turn to the specific case of the Oslo II water accords. These accords, as we have already seen, allowed Israel to free itself of direct responsibility for Palestinian water infrastructures and Palestinian water supplies (these responsibilities being passed on to the newly established PWA and international donors), and also enabled Israel to transfer the West Bank Water Department's debts to the Palestinian Authority. At the same time, the accords gave Israel licence to continue exercising sole control of the West Bank's water resources, granted it the right to veto any proposed Palestinian infrastructure developments, and enabled it to carry on feeding its settler population with plentiful, subsidised supplies of water. Under Oslo II, the Palestinians became responsible for local water supply management, while Israel remained in control of the water resources. All of this was very much in keeping with the terms of the 1989 Baker–Shamir–Peres Plan: the PA was granted 'autonomy' over the 'daily life' business of administering local water supplies; Israel on the other hand remained in overall control.

Most Palestinian water experts had understandably wanted something quite different from the Oslo II negotiations, arguing above all that any agreement should address questions of ownership and rights. Palestinian water experts have argued for a long time that Israel's exploitation of the West Bank's waters is in contravention of the 1907 Hague Regulations on military occupation, and also that the Palestinian right to sovereign territorial control over the West Bank implies at the same time rights to the water lying within it. Palestinian water experts thus typically claim for the Palestinians 'absolute sovereignty over all Eastern Aquifer resources, as this aquifer is completely located beneath the West Bank and is not a shared resource'; they claim joint rights to the trans-boundary Western and North-eastern aquifers, maintaining that Israel should have access only to a limited proportion of their waters; and they also claim some rights to the Jordan River, arguing that Israel's diversion of water from the Sea of Galilee prevents that water from reaching the West Bank, where it would otherwise be used by Palestinians.[37] It might be noted in passing that these arguments about the West Bank aquifers are inevitably premised on the positing of genuine natural objects called the Eastern Aquifer, Western Aquifer and so on – of objects which, as we have already seen, are in some sense social constructs. In any case, throughout their water negotiations with Israel, the Palestinians' main recourse has been to principles of international water law.[38]

Against this Palestinian position, Israeli experts and water officials have typically argued that they themselves enjoy significant 'prior use' rights to the West Bank aquifers. Israel submits that just as Egypt has legitimate rights to most of the water of the Nile River by virtue of the fact that it was using them prior to Sudan, Ethiopia and so on, so too does Israel have rights to the shared West Bank aquifers by virtue of the fact that it has been exploiting these since at least the 1950s, at a time when these waters were barely being exploited from within the Jordanian-controlled West Bank. Israelis argue that this confers on them rights to the Western and North-eastern aquifers that are equal to or greater than those of the Palestinians.[39] Israel is determined to ensure that these rights – or at least these supplies – are not threatened, and hence it refuses to countenance further Palestinian development of the Western and North-eastern aquifers, and also insists on maintaining control of these aquifers' recharge areas.[40] Israel is also unwilling to recognise Palestinian rights to the Jordan River. By contrast, Israel has been quite willing, and even keen, to defer responsibility for Gaza's coastal aquifer, simply because increased abstraction from this could not affect Israel's own supplies – this of course explains why the PA became solely responsible for operating, managing and developing water resources and systems in the autonomous areas of Gaza, but was denied these rights when it came to the West Bank.

Beyond contesting Palestinian interpretations of international water law, Israel has also sought within negotiations to marginalise and defer all discussion of ownership and water rights. The words of former Water Commissioner Gideon Tsur are typical of this position, 'the issue of water rights is secondary,' he says. 'Not one cubic metre of water has been created by a declaration of rights. And I'd like to see us begin to talk about… creating new quantities of water.'[41] In keeping with this line of argument, Israel refused to discuss rights issues within the Madrid multilateral negotiations on water, maintaining that these talks provided a forum for exploring technical rather than political matters.[42] Equally in its bilateral negotiations with the Palestinians, Israel has consistently sought to defer rights issues to final status talks, arguing that the interim period of the Oslo process should be concerned with practicalities, not principles.[43] During the early Oslo negotiations, Israel sought to focus on Palestinian 'needs' rather than rights, and to emphasise the importance not of transferring resources to the Palestinians, but of increasing their supplies. 'Let's be realistic,' Shimon Peres reportedly once said to Yasser Arafat. 'We won't take a drop of water from you [sic!]. I suggest that all the water you now have will remain at your disposal, and we'll try to help you out by finding new water.'[44] The Palestinian 'needs' that the Israelis were suddenly so concerned about were, it should be noted, purely 'humanitarian'; Israel has consistently made it clear that it will not divert water from its own agricultural sector for Palestinian agricultural purposes.[45]

Finally, Israeli leaders were motivated by a desire to defer responsibility for the terrible state of Palestinian water supplies, and to do this without losing control over vital resources. A television documentary broadcast during August 1995 had drawn close attention to the dire summer water shortages in Hebron, contrasting these with scenes of plenty from nearby Kiryat Arba. These inequalities received wide media coverage and also condemnation within Israel and, on the strength of this, Yitzhak Rabin responded not only by deciding to start trucking water immediately to Hebron, but also by making the strategic decision to separate Israeli and Palestinian water networks within the West Bank to as great a degree as possible.[46] Most Palestinian experts had long been arguing for separation of water supplies, this being seen as a necessary corollary of self-rule. Now Israel wanted separation too. It was on the strength of this convergence of interests that the co-ordinated water supply management system enshrined within the Oslo II Agreement was constructed.

There were understandably some significant disputes amongst Israeli experts and policy-makers regarding the water provisions of the Oslo II Agreement. To the right of the Israeli political spectrum, the water accords were denounced, as we have already seen, as a threat to the country's vital national interests. From elsewhere, a group of Israeli water professionals called, in association with their Palestinian counterparts, for more fully 'joint' (rather than merely 'co-ordinated') management of the West Bank's resources and systems.[47] Within Rabin's government, moreover, there were tensions between Peres's liberal-civilian team who, during the secret Oslo negotiations had implicitly recognised Palestinian water rights, and Rabin's more hawkish and national security-oriented Defence Ministry officials, who believed that any explicit recognition of Palestinian water rights would but create problems during future negotiations. These tensions were representative of the broader conflict between Israel's national security establishment and its new liberalising elites. Nonetheless, they had been largely resolved – or more accurately, had been controlled and structured – before Israel came to the Oslo II negotiations with the Palestinians, with Peres and his staff taking charge of the less weighty, multilateral Arab–Israeli talks, and Rabin's men overseeing the bilateral water negotiations with the Palestinians.[48]

Indicatively, the Israeli team at the Oslo II negotiations was led by Noah Kinnarty of the Ministry of Defence, a hawkish military man in the style of Rabin, who prides himself as, and is renowned as being, a 'very tough negotiator' with 'a lot of experience of negotiation with the Arabs'.[49] Kinnarty was at the helm of a core team of half a dozen negotiators, which in turn drew upon the advice of a back-up steering committee, and also upon the expertise of Tahal, Mekorot, the IHS and the Water Commission as and when necessary.[50] Kinnarty himself reported directly to Rabin, with whom he had established precisely laid-out 'red lines' for the negotiations.[51] Thus while the Israeli negotiating position reflected the hawkish views of Kinnarty and Rabin, it nevertheless also benefited from the expert advice of Israel's state water institutions and leading water experts.

In contrast with this, the Palestinian team to the Oslo II water negotiations was riven with dispute and uncertainty, both in terms of negotiating positions and with regard to decision-making authority. The team consisted of six men: Marwan Haddad, a university professor who was nominally head of the delegation; Abdelkarim Asa'd, manager of a Palestinian water supply utility; Taher Nassereddin and Mustapha Nuseibi, both senior figures in the West Bank Water Department; and Nabil Sharif and Fadel Qawash, two close supporters of Arafat who had only recently arrived in the Territories from Tunis, and who were later to become head and deputy head of the PWA.[52] In addition, Arafat's chief negotiator, Abu Ala, carried some responsibility for overseeing the Palestinian negotiating team and position.[53] However, the various negotiators' powers and responsibilities were not firmly established, and the team as a whole lacked clearly agreed aims and objectives.[54] On certain issues they held quite contradictory views, with Marwan Haddad favouring the creation of a joint management system, for instance, and others supporting institutional separation.[55] Moreover, given that there were no national Palestinian water institutions in existence, the team understandably had neither firm data nor thoroughly worked plans and projections that it could call upon. To make matters still worse, two of the six were still receiving their paycheques from the Israeli Civil Administration, and another two were participating in the negotiations by virtue of their closeness to Arafat, not because of any detailed knowledge of the water situation in the Territories. The structural differences between the Israeli and Palestinian negotiating teams were stark indeed.

Negotiations took place during the summer of 1995 and, for several months went without agreement on even a single paragraph. The sticking point here was the Palestinians' insistence – and on this there was unanimity – that Israel recognise their water rights. Kinnarty refused to do this, and the negotiations consequently hit a stalemate which was only overcome through higher level talks between Abu Ala and the Israeli Agriculture Minister Ya'akov Tsur, these talks leading to the pre-signing of the paragraph recognising 'Palestinian water rights in the

West Bank,' which was later to become the first paragraph of the Oslo II water accords.[56] Thereafter, attention turned to the remaining and more technical issues. There was a clear danger that the water negotiations were going to hold up the Oslo II Agreement as a whole, to the extent that Arafat was pressurised for the Palestinians to make concessions. Suddenly the Palestinian team was effectively dissolved, such that one man was left to negotiate with the Israeli machine. This was not Marwan Haddad, who was nominally at the head of the team, but Nabil Sharif, who had expertise neither in water issues, nor in the local situation in the Territories, but who had been close to Arafat ever since the mid-1960s, and is still to this day one of Arafat's most trusted associates.[57] Israeli pressure, combined with Arafat's penchant for switching around his negotiators in an *ad hoc* and thoroughly patrimonial fashion, had ensured that the final leg of the Oslo II water negotiations would be conducted between a relatively uninformed individual and an Israeli administrative state machine that had a full panoply of plans, policy documents, facts and figures at its disposal. Faced with an Israeli delegation which would not go beyond its clearly demarcated red lines, Nabil Sharif was in an unenviable position, and it is thus hardly surprising that he ended up signing an agreement that so overwhelmingly reflected Israeli interests. As one Palestinian water expert bitterly observed, 'the Palestinian negotiators ... didn't read the Article before signing, otherwise they don't sign ... the technical people, they didn't sign any paper to the Israelis. This is signed by politicians, even the pre-signature. It's done by politicians, and this is the result' – a dressing up of domination as co-operation.[58]

Ridiculous as this situation was, it was not atypical within the Oslo negotiations. In virtually every sphere one had a well-co-ordinated Israeli team with well-worked negotiating positions and plenty of facts, files and institutional back-up; and a barely organised Palestinian team, lacking clearly developed proposals and strategies, and lacking a stable decision-making hierarchy. To give just one further example, the PLO negotiated the Declaration of Principles without the aid of a legal advisor![59] Accounting for such PLO inadequacies, Norman Finkelstein contends that 'the PLO's capitulation at Oslo did not result from political ineptitude ... The problem was, they had no power.'[60] Yet this assessment is surely not quite correct, since it rests upon a false opposition: *pace* Finkelstein, the PLO was inept in negotiations with Israel, but this was in large part because it lacked its institutional and decision-making structures, and its own consensually accepted plans and policy documents, and had little means of producing them. This in turn was because, throughout the occupation period, Israel had worked to retard the development of national Palestinian institutions in the Territories; because the outside PLO had itself been more oriented towards mobilisation and armed struggle than institution-building; and because, while there were plenty of Palestinian NGOs and technical experts within the Territories, these were faction-based, often in competition with one another, and indeed were often excluded from (or reluctant to participate in) the negotiations.[61] By contrast with the Israeli system, then, there was little basis within the politically fractured Palestinian arena for technical aptitude, or for the development of consensually agreed plans, proposals and policy documents (these issues will be discussed at greater length in the next chapter). Unlike Israel, the PLO had neither macro-political support from the US, nor a brutal military machine, nor institutional structures and procedures for ordering decision-making and constructing socially stable truths. Edward Said's critique of the PLO is in this regard spot-on: the Israelis, he writes – in terms that betray the influence of Foucault's observation that 'discipline is a political anatomy of detail'[62] – that they 'had the plans, the territory, the maps, the settlements, the roads: we had the wish for autonomy and Israeli withdrawal, with no details and no power to change anything very much. Needed: a discipline of detail.' It was only in Israel's power, for instance, to define there as being sufficient additional potential in the Eastern Aquifer to satisfy Palestinian needs well beyond the interim period – for despite the fact that this claim was highly uncertain, and despite the fact that most of the hydrological data on which it was based had initially been collated by Palestinian technicians at the West Bank Water Department, the Palestinian Oslo II negotiators had no means of assessing or contesting its veracity. Nabil Sharif certainly did not, and it was largely for this reason that Oslo II ended up defining the Eastern Aquifer as having an additional potential of 78 mcmy, and stipulating that this would be the chief resource for developing new Palestinian water supplies.

Having agreed to these terms, the PA and PWA were from then onwards bound by them. As with the Declaration of Principles, once the new, co-ordinated management system was accepted there was little that could be done to restructure or redirect relations. It was for this reason that the PWA became willing to license the construction of new supply lines to Israeli settlements in the West Bank from inside the Green Line. Faced with recurrent Israeli vetoes on its own infrastructure development projects, the PWA in the end felt compelled to grant such permission, and to reach a *modus vivendi* with Israel – since without this it would not have received approval for its own infrastructure developments. In justifying its proposals, Israeli officials have stated that once its settlements are sufficiently linked with lines from within the Green Line, they will no longer need to be supplied from wells within the West Bank, whereupon these Mekorot-controlled wells will be transferred to the PA in accordance with the stipulation that all systems 'related solely to Palestinians' should be maintained by the Palestinians themselves.[63] There is so far no evidence of any such transfer coming to pass.

Cecilia Albin has argued in a detailed analysis of the Oslo II negotiations that 'power inequality alone cannot explain the Israel–PLO interim talks' and 'did not drive the water talks'. 'Both parties were determined to reach an agreement and had much to lose from failing to do so. This meant that Israel had to take into account Palestinian notions of a reasonable solution.' Hence 'Israeli power was restrained by the need to give the Palestinians a fair share of benefits' – that is, a recognition of their water rights.[64] Albin reaches this conclusion, however, only by eliding the fact that the Palestinian delegation evaporated into thin air during the final stage of negotiations, by ignoring the extent to which the supposedly new JWC structures represented a continuation of the system that had prevailed under occupation, and by focusing almost entirely on the question of water 'rights'. Yes, certain Palestinian rights were recognised in the Oslo II water accords, and this may prove important in future negotiations. Nonetheless, to ascertain the relative significance of 'power' and 'fairness' in the Oslo II water talks, we need to focus primarily not on language and discourse – since negotiators will always seek to present their agreements as either in the national interest, or for the common good, or more likely both – but rather on the extent to which powers, freedoms and responsibilities have been structured and divided. And if one considers the Oslo II water accords in this fashion, with an eye to material and institutional changes, then it becomes very hard to see any evidence of 'fairness' considerations having had any impact whatsoever.

CHAPTER 7

Administering Water Under Oslo

In May 1994, Israel and the PLO reached agreement on the precise terms of the initial phase of Palestinian autonomy, and before long the PA was formally established in Gaza and Jericho. By early 1996, the PA's authority had been extended to the major Palestinian population centres across the West Bank. The Authority faced an unenviable task. Not only was it charged with being Israel's enforcer in the Occupied Territories; it would also be responsible for administering services to the Palestinian population, and for rehabilitating and developing public institutions and infrastructures across the West Bank and Gaza. At the same time, the PA had itself to undergo a rapid phase of institution-building. Prior to May 1994, the PA had no ministries, had in waiting only the nucleus of a police force, and indeed did not formally exist at all. Arafat and the PLO leadership would have to oversee the formation, development and functioning of a proto-state, the Palestinian Authority.

This chapter analyses these processes in relation to the Palestinian water crisis, my key aim being to assess why the southern West Bank was still suffering from water shortages several years into the Oslo process. Part of the reason for this has already been discussed – namely that the Oslo II Accords enabled Israel to continue controlling virtually all shared water resources, licensed it to continue discriminating against Palestinians in its water supply practices, and gave it a veto over all Palestinian infrastructure developments. The main difference now was that all of this was being done with official PLO consent. Nonetheless, this does not fully explain why during 1998, 1999 and 2000, well before the region descended into violence, so many Palestinians were still facing water shortages. The PWA had been granted in Oslo II the right to develop significant new water supplies from the Eastern Aquifer, and despite problems during 1997 under Netanyahu, was nonetheless receiving permissions through the JWC to do so. Thus to explain the ongoing water crisis – the fact that most areas of Hebron received piped water for just one day every 20 during summers 1998 and 1999, for example, or that the village of Quasiba received no supplies for five months – to explain this we need also to consider the internal problems faced by the PA and PWA in managing the water sector, and in constructing their new infrastructures.

In previous chapters we have seen that the specific character of the Zionist colonial encounter with the Palestinians led to the emergence of state and pre-state institutions that were very powerful in relation to the society, economy and territory under their control. Israel became a quintessentially 'strong state', one that did not have to battle against local 'strongmen' contesting the power and legitimacy of the central state apparatus.[1] On the back of this, a powerful complex of central water institutions emerged, and all water resources were brought under state control. The contrast with the Palestinian case is stark indeed. Under Oslo, the West Bank Water Department and the Palestinian Water Authority (PWA) have had immense difficulties in controlling the actions of local Palestinian municipalities, village councils and even individual actors, all of whom have had their own local interests and agendas to protect. The PWA has also been weak in relation to international donors and their private sector contractors, such that the latter's interests, rather than those of the PWA, have often prevailed. These institutional weaknesses, I argue, have greatly complicated the process of implementing the Oslo II water accords, this being part of the reason why places such as Hebron and Quasiba continued to face water shortages.

I begin by considering in broad terms the nature of the Palestinian political system under autonomy. Thereafter we move on to the water sector, and examine a series of cases of PA-society and PA-international donor relations – these relating in turn to the field of water planning, to the administration and governance of existing water supply systems, and to the construction of those new infrastructures made possible by the Oslo II Agreement. Each of these mini-case studies illustrates, in different ways, the troubles that the central PA water institutions have had in attempting to improve the water situation in the Palestinian territories. Moreover, each of these cases points once again to the intricate relations that exist between politics, techniques and technologies – with most of the conflicts charted here arising not over water resources directly, but over plans and policies, wells and pipelines, and even water valves.

PALESTINIAN POLITICS AND SOCIETY UNDER OSLO

It did not take long for the Palestinian political system under Oslo to take shape. Within weeks of Yasser Arafat's arrival in Gaza, he had created a political system in which financial resources and decision-making power were concentrated in his own hands; in which formal structures were routinely bypassed and ignored; and in which power was principally exercised instead through informal patronage networks, and through his control of police and security services.[2] These developments are often explained as resultant from Arafat's personal style of rule, and there is doubtless a great deal to such interpretations: we have already seen, after all, that Arafat's Oslo II water negotiating team operated in a haphazard and personalistic fashion, and lacked formal decision-making structures. Nonetheless, the nature of the Palestinian political system under Oslo can, in my view, be more adequately explained in structural terms, as the determined product of state-society relations under the occupation, of the PLO's history of armed struggle, and of the constraints imposed by the Oslo agreements. Given this, we must start by considering the internal Palestinian political environment that Arafat inherited, and into which the PA was born.

The political environment that the PA inherited was riddled with conflict between PLO factions, and between them and the non-PLO Islamist groups, Hamas and Islamic Jihad. These various conflicts had grown throughout the early 1990s, to the extent that there had been street battles in Gaza between supporters of Fatah (the PLO's dominant faction, led by Arafat) and those of Hamas.[3] While the latter was an increasing threat to Israel, it was also increasingly a threat to the internal ascendancy of the PLO, and a challenge to Fatah in particular. Arafat's capitulation in the Oslo agreements understandably exacerbated many of these internal Palestinian conflicts, and at the same time engendered new ones. Hamas and Islamic Jihad rejected the agreement, as did the Marxist and Syrian-backed Popular and Democratic Fronts for the Liberation of Palestine (PFLP and DFLP). Moreover, across the Territories, Arafat's Fatah faction was itself divided over Oslo. This in turn had effects on the ground, such that across the West Bank, and even more so in Gaza, the period between the signing of the Declaration of Principles and the establishment of the PA was marked by internecine fighting and civil disintegration.[4]

At the time there were few national institutions – a historic consequence of Israeli, Jordanian and also diaspora-based PLO attempts to obstruct institution-building within the Territories (the PLO 'outside' being motivated in this by a desire to maintain political hegemony over the 'inside' wing of the national movement).[5] Moreover, all institutions were tied to one or other of the factions – this even applied, for instance, to trade unions.[6] Thus the Palestinian political arena (and civil society more broadly) was deeply politicised, being fractured along factional lines. During the early years of the intifada, the Unified National Leadership had functioned as an inside-based and non-factional national institution, responsible for articulating national strategy and mobilising resistance against Israel, but this had become increasingly subordinated to the Tunis-based PLO and thus was, by 1993, largely inoperative.[7] In this fragmented political environment, power lay primarily with the factions and their armed wings, secondarily with local municipalities and village councils, and thirdly with non-governmental organisations. In the West Bank, municipalities had been given relatively free rein by the Israeli authorities, and prior to that by the Jordanian authorities, in administering local Palestinian affairs (we have already seen, for instance, that municipalities were responsible for supplying water within internal networks, and for collecting payments to be made to the West Bank Water Department). In the absence of a state or of over-arching national-level institutions, these local municipalities had amassed significant sub-national powers which, come the arrival of the PA, they would no doubt seek to defend. Likewise, thousands of Palestinian NGOs had been established during the 1980s, these playing a vital role – given the absence of a state – in delivering health, education and other community services, and in serving as forums for international advocacy and internal political mobilisation.[8] These NGOs were not co-ordinated under a single umbrella, however, and indeed were commonly tied to individual factions, and especially to the Palestinian left (PCP, PFLP and DFLP).[9] They thus served as organisational and power bases for their factions, and also for the educated 'insider' political elites who headed most of them. Like the municipalities, these NGOs had important powers and responsibilities, which they would no doubt be loath to sacrifice to Arafat and the PA. All in all, Arafat inherited a factionalised and highly fragmented political arena.

In addition to this, during the early 1990s the Territories were also in the midst of a serious economic downturn. Most Palestinian institutions in the West Bank and Gaza were dependent on outside-controlled PLO funds, and given the organisation's financial crisis – a result of its Gulf War support for Iraq – all of these NGOs, universities, health care centres and so on were suffering serious cut-backs. In addition, Israel's closure of the West Bank and Gaza in March 1993 had produced a large rise in unemployment within the Territories, and consequent economic hardship. The PA would also, of course, inherit social and economic infrastructures that bore the marks of 27 years of Israeli neglect and 'de-development'.[10]

If these economic and political problems did not pose enough problems, Arafat and the PA would also be severely constrained by the terms of the Oslo Accords. These, as already noted, granted the Authority power only in those areas and spheres where Israel was keen to cede responsibility. The PA's primary responsibility, as far as Israel was concerned, would be to police and suppress the Palestinian opposition, especially Hamas and Islamic Jihad. Beyond this, the PA would be responsible for governing and providing local services to Palestinians across the West Bank and Gaza. It would have to do so, however, while exercising only limited territorial control within the Territories, and while lacking any degree of economic independence from Israel. The PA would also inevitably have great difficulty in raising capital for public services, since the economy was in such decline, and since the level of tax-evasion was so high (a consequence of the intifada, when Palestinians were urged by their leadership to refuse to pay Israel's exorbitant taxes).[11] In short, the terms of the Oslo Accords – given also that Palestinian civil society was so fragmented and conflict-ridden – would inevitably structure and constrain the process of constructing the Palestinian proto-state.

The character of the PA under Oslo can be seen as a strategic response to these structural constraints. Desperate to assert and maintain control over the Gaza Strip and, somewhat later, the West Bank, Arafat and the PA leadership came to rely above all on trusted outsider associates, on factional loyalties, on powers of patronage, and on the police and security services. Arafat's entry into Gaza was akin to a 'military takeover'.[12] He himself was preceded by a 7000-strong police force, formed from units of the Palestinian Liberation Army (PLA) which had previously been based in Iraq, Libya and elsewhere in the Arab world, and which became the core of the PA police force in the Gaza Strip and Jericho. Local Fatah activists were also recruited, both into the civil police and into the various security agencies (where they predominate). As noted earlier, under Oslo there were as many as 14 extra-legal security agencies. All of them answered to Arafat, and all were in competition with one another, thus enabling the PLO Chairman to maintain firm control of his new fiefdoms and prevent rival power centres from emerging.[13]

Throughout the PA system, key posts were given to Fatah supporters, such that Arafat's can quite reasonably be characterised as a one-party regime. The most important positions within both the police forces and PA ministries were given to PLO leaders and functionaries from Tunis, presumably judged by Arafat to be more loyal than insiders, and less of a potential threat to his authority, since unlike insider leaders they lacked domestic constituencies. Other key posts were given to members of the traditional Palestinian landowning elite who, unlike the students, workers and prisoners who grew to prominence during the intifada, also lack strong domestic constituencies.[14] Fatah supporters were handed most of the key posts even on lower rungs of the administrative ladder. Insider activists were and remain under-represented within the PA system, except at lower and more functional levels of the PA bureaucracy.

The PA has a full set of ministries, authorities and other such agencies. However, policy decisions are taken largely by Presidential fiat since, notwithstanding the formally existing institutional structures, Arafat pronounces 'on everything from the minutiae of Gaza's sewerage system to the future of the Palestinian women's movement'.[15] Despite the existence of the elected Palestinian Legislative Council, decision-making power continues to lie overwhelmingly with the executive. The PLC's recommendations and censures are often ignored by Arafat, as was the case for instance when the Council demanded the sacking of several key ministers on corruption charges, and attempted to pass a motion of no confidence in his cabinet.[16] Rumours and often evidence of corruption run rife through the upper echelons of the PA. The Palestinian Authority, in short, is a strongly neo-patrimonial regime in which, despite formal adherence to institutional norms, decision-making is personalised and concentrated in Yasser Arafat's own hands.[17]

Israel and the international donor community have both had a hand in these developments, each providing sources of funding upon which Arafat's powers of patronage depend. Until recently, the income tax deducted by Israel from the wages of Palestinian day labourers was forwarded not into an official PA bank account, but into one of Arafat's personal ones: this arrangement changed during early 2000 on the back of IMF pressure, no doubt to Arafat's disappointment.[18] As for the donors, while they have been as keen as ever to construct transparent aid disbursement systems, and to promote norms of 'good governance' within the PA, the evidence is that their aid moneys have had decidedly mixed effects upon Palestinian politics and society.[19] Initially, the aid process was intended to prioritise long-term projects, and to promote the sustainable development of the PA and Palestinian economy. However, in the wake of Israeli closures during 1994 and 1995 and the consequently high levels of Palestinian unemployment, a large portion of the early funds were diverted to short-term employment generation schemes, in a desperate effort to 'buy peace' and prevent the collapse of the Oslo process.[20] Other funds were used to support the Authority in employing perhaps 100,000 people within its bloated ministries and police and security services.[21] In spite of the efforts of donors, funds have often found their way into the private coffers of PA ministers and functionaries, providing sources for patronage. Moreover, the massive reliance on donors, and problems of donor co-ordination and competition, have in various ways complicated and retarded the institutional development of the PA. Thus while these development assistance funds have doubtless been central to the process of constructing the PA proto-state – many PA ministries would barely exist without such funds – they have at the same time fed prevailing anti-institutional dynamics.

The establishment of this PA system inevitably led to tensions both with municipalities and with the NGOs. Mayors were appointed by Arafat in 1996, and these remain in position: municipal elections have not been licensed by Arafat, presumably since these would effect a lessening of his control over the municipalities. As for the NGOs, their international funding predictably decreased after 1993, with donors redirecting most of their development assistance to the newly formed PA.[22] Even so, the PA has endeavoured both to control their international funds, and to regulate their activity.[23] There has often been conflict between the NGOs and the PA, often for inter-factional reasons, with Arafat once pronouncing, for instance, that NGO activists were 'all communists'.[24] For their part, many of the NGOs are still today bigger than their respective ministries, and most of the insider and educated NGO elites have remained outside the PA, often hostile towards, or at best ambivalent about, the PA and the Oslo process. The PA ministries have suffered as a result from a dearth of local Palestinian expertise.

THE CENTRAL PALESTINIAN WATER INSTITUTIONS: AN OVERVIEW

With this broad picture established, we can now turn more specifically to the Palestinian water institutions' attempts at governing water systems and supplies. We have already seen that, since 1995, the Palestinian water sector has officially been under the authority of the PWA, but that, in the West Bank at least, the Water Department is still in many respects the more important institution, with the former being responsible for overseeing internationally funded projects, and the latter being charged with everyday water management. The West Bank Water Department, as we have seen, is still officially part of the Israeli Civil Administration. The PWA, by contrast, is nominally independent – though in practice very much a donor construct. As of summer 2002, only five of the PWA's West Bank and Gaza personnel were on the PA payroll, the remainder being employed on a project basis, principally by the US Agency for International Development, the United Nations Development Programme, the World Bank, and the French and Norwegian development agencies.[25] Since 1996, the PWA's offices have been funded largely by a Norwegian capacity-building and institutional development project, with UNDP also until recently providing institutional support.[26] The World Bank, UNDP and others have financed and been instrumental in the production of facility plans, master plans and strategic plans for the Palestinian water sector. USAID, in particular, has been funding large-scale infrastructure development work in both the northern and southern West Bank, and has employed a large number of international and local contractors within (or in support of) the PWA. These various sources of funding have been vital to the development and work of the PWA. Nonetheless, and as we shall see, this international development assistance has inevitably generated conflicts within the PWA, and has to a certain degree complicated its institutional development.

The PWA is more fully developed as an institution than most other PA ministries or agencies, in large part because it has been the recipient of a disproportionate amount of donor funding, and in part too because it is headed by one of Arafat's most trusted confidants, in whose work Arafat does not greatly interfere. However, like every other corner of the PA system, the PWA is characterised by neo-patrimonial dynamics, many of these being exacerbated both by the structural constraints of the Oslo Accords, and by the predominance of international donors and donor monies. Typically, as already mentioned, the PWA is headed by two outsiders: Nabil Sharif is the head of the PWA, and based in Gaza, while Fadel Qawash is his deputy, and based in the West Bank's administrative capital, Ramallah. Typically also, this geographic split – often exacerbated by Israeli closure – has created problems of control and co-ordination, with Fadel Qawash as a result generally acting as *de facto* head of the PWA in the West Bank.[27] Within the West Bank itself, the PWA has been located in a number of different sites, with each of these having its own head, and each primarily receiving its funding from its own international donor. As a corollary of this, the heads of these different sites have come to voice, to a certain extent, the concerns of their respective donor backers, such that contending donor interests, priorities and strategies have been internalised within the PWA. A related problem is that, because of the presence of so many donor organisations, the PWA has a variety of different wage structures: as of 1999, there were four such wage structures in the West Bank alone.[28] Understandably given all this, the PWA has often not functioned along clearly structured institutional lines.

Besides these internal problems, two other introductory points need to be made. First, and as with the PA more broadly, the PWA has faced difficulties in controlling municipal authorities, and in centralising administrative power, and has thus come to rely in large measure on informal and factional means of extending its administrative reach. In some cases it has been successful in this: within Bethlehem, for instance, the director of the Water Supply and Sewerage Authority was sacked and jailed in 1996 on (quite possibly spurious) corruption charges, to be replaced by a local Fatah activist, who happened to be a school teacher rather than a water expert or official.[29] The PWA has since attained firm control over the WSSA, such that donor work in Bethlehem has not been complicated by disputes between central and municipal authorities. In other cases, though, the PWA and PA have not managed to tame the municipalities: Hebron municipality, for instance, has sought to defend its important water and other resources, and this has created multiple difficulties both for the PWA, and for donors, as we shall see later.

Finally, the PWA has suffered from a deficiency of water experts. The Palestinian water scene is often praised by international experts as comparing well in its expertise with neighbouring Arab states. Unfortunately, however, most of these experts remain outside the PWA, in the NGO sector. Several water-related Palestinian NGOs and university-based institutes were established in the Territories during the 1980s and early 1990s, these generally undertaking small-scale water development projects, and research on local water (and related environmental and agricultural) issues. Since Oslo, these NGOs and institutes have necessarily had to adapt their roles to meet the new demands of the PA and international donors. Nevertheless, none of the 'institutional entrepreneurs' who founded water-related NGOs during the 1980s, have left for the PWA, partly because this affords them greater institutional and political independence, but no doubt also for financial reasons.[30]

With these general points in mind, we can now consider in greater detail the problems faced by the central Palestinian water institutions in three areas: administration and planning, the management of water supplies and systems, and the construction of new supply infrastructures.

The second of these issues requires us to focus on the West Bank Water Department; the first and last, by contrast, centre primarily on the institutional functioning of the PWA.

STRUCTURING ADMINISTRATION, PLANNING FOR PALESTINE

Since its inception, much of the PWA's work has focused on institutional development, capacity-building and national planning, all of these tasks being judged necessary, especially by donors, in developing a modern and well organised water sector. In the Palestinian context, however, all such attempts have been beset by difficulties.

The PWA formally has four separate departments, these being responsible for Water Resources and Planning; licensing, tariffs and consumer affairs (the Regulatory Department); research and water sector capacity-building (the Technical Department); and management of PWA personnel and finances (the Administrative Department). Developed through the Norwegian-funded institution-building project mentioned above, this institutional structure is codified within the PWA's Internal Regulations of 1996.[31] This formal structure does not, however, correspond to the actual workings of the PWA, as can be seen in relation to the Water Resources and Planning Department, and the Regulatory Department. With regard to the former, the PWA is formally responsible for '[i]nspecting, supervising and administering all Water Resources and their different uses'.[32] The reality, however, is that it is the Water Department that inspects and administers all water resources through the JSETs teams, and Israel that controls all water resources. The PWA's Water Resources and Planning Department has, since 1996, been collating data gathered by the Water Department into a data bank, but in truth has very little administrative control of water resources. As for planning, the Water Resources and Planning Department has indeed undertaken some strategic planning, but so also have people working elsewhere within the PWA, funded by their own and different donors. For its part, while the Regulatory Department has been developing a tariff system in accordance with its mandate, little of this has or could have been implemented, since water prices from Israeli-controlled sources are all set by Mekorot.[33] Most strikingly of all, most of the PWA's work involves co-ordinating donor-funded infrastructure projects, yet this work does not fit anywhere into its formal institutional structures.

A key reason for this is that the PWA's internal structure and provisional water policy represent administrative ideals which will hopefully be realised at some future date, when agreements with Israel confer greater independence and power on the PA in governing water resources, systems and supplies. Whether future agreements will in fact do this is very much an open question. In the long-term, the PWA aims to be a planning and regulatory body akin to the Water Commission in Israel, involved neither in managing infrastructures, nor in distributing supplies.[34] However, given the absence of any other national water institutions, the PWA has inevitably had to assume, in the immediate-term, a major role in co-ordinating donor-led infrastructure projects, and thus necessarily engages in work that is not mirrored in its formal administrative structures.[35] Given that the PWA's formal structure refers to a hoped-for future rather than to the present, the agency in practice has no administrative structure whatsoever.

This state of affairs might be viewed as inevitable given the constraints and demands of the Oslo process. Nonetheless, one may well wonder why the PWA has taken the effort to formulate structures – but also policies, plans, procedures, and so on – which cannot yet, and might never, be realised. Part of the answer here evidently lies in relation to donors. The PWA developed its provisional water policy at the request of the World Bank, as condition of the Bank's support for a major water and wastewater project in the Gaza Strip.[36] Likewise, we have already seen that the PWA's formal structure was created as part of a Norwegian-funded project (another institution-building project, funded by Denmark, proposed a quite contrary structure for the PWA; this project and its findings were sidelined for internal political reasons).[37] Palestinian water experts, and even PWA officials, often charge that donors are obsessed with plans and structures, frequently to the detriment of the much more needy and practical work of constructing infrastructures, and developing new supplies.[38] Given the constraints and uncertainties faced by the PWA, it is indeed hard to see what the value has been in developing an institutional veneer that bears so little relation to reality – other, that is, than to spend donor money, and to provide employment for Palestinians and for foreign consultants.

Many of the same points can be made in relation to water planning. Since the formation of the PWA in 1995, countless plans have been produced for the Palestinian water sector. Some of these have been facility plans, detailing proposals for short-term and long-term infrastructure developments; some have been district-level plans, setting out projected demand and supply requirements within particular locales;

some have been regional plans, detailing the same in relation to the two Palestinian 'regions' of the West Bank and Gaza; others have been master plans or strategic plans for Palestine as a whole (i.e. for the Gaza Strip and West Bank, including East Jerusalem); and still others have been multilateral plans, produced in association with Israel, Jordan and other states, and aiming to develop proposals for region-wide supply and demand management.[39] These plans have all been funded by international donors; and they have all been produced by international consultants, generally in collaboration with the PWA, but in some instances with other ministries, such as the Ministry of Planning and International Co-operation.

These planning exercises have all been beset by institutional and political difficulties. MOPIC's 'Water and Wastewater Regional Plan for the West Bank', for example, was a site of conflict between MOPIC and the PWA, and between them and the German and Palestinian consultants, over a wide range of technical and political issues.[40] PWA officials objected to MOPIC undertaking such a plan, and averred that they lacked the expertise to do so effectively, while MOPIC officials, for their part, charged that the PWA had withheld information from both themselves and the consultants. In part as a result of the dearth of contact between consultants and the PWA, but also because consultants typically have the barest understanding of the Palestinian water sector, much of the plan's factual content was poor, and many of its assumptions were misplaced (for instance the plan was developed in complete ignorance not only of existing wells in the West Bank, but also of the ongoing USAID-funded infrastructure work in the southern West Bank). Moreover, a large number of the consultants' ideas were vetoed on explicitly political grounds. PWA and MOPIC officials were unwilling to license proposals that, if Israel were to learn of them, might conceivably detract from the Palestinian negotiating position (they objected, for instance, to the contractors' assumption that existing Israeli extraction in the West Bank would in future remain at present-day levels). PWA officials even charged that the plan contravened PWA policy. They demanded that the contractors substantially revise their plan, to which the latter responded that substantial revisions would be impossible since (at the moment when the plan was presented) their contract had only two days to run! A plan had been produced towards which PWA and to a lesser extent MOPIC officials were deeply hostile, one that will undoubtedly never be implemented by the PWA.

While this MOPIC plan was the source of an unusual degree of conflict – plans produced for the PWA have not suffered from such inter-agency rivalry and enmity – it does nevertheless illustrate the enormous institutional difficulties that have been faced by the PWA, especially in relation to donors. Consultants typically have only a threadbare knowledge of the local Palestinian water situation, bringing instead a non-area specific expertise to the planning process. This lack of local expertise inevitably means that Palestinian officials are forced to do a great deal of data-gathering 'donkey work' for their international consultants, while the consultants themselves do the expert work of composing final planning documents. Besides causing some resentment, the predictable result has invariably been that the PA's water sector plans have reflected consultants' judgements above all, with the knock-on effect that PWA officials have at times disagreed with the content of their own plans, and have often felt a lack of 'ownership' of them.[41] Equally, because successive plans have been conducted by very different groups of consultants, these plans have been marked both by a great deal of duplication, and by contradictory information (see for instance the maps shown together in Figure 5.1, all produced within donor plans). The production of so many water plans, and the presence of so many different international consultants, has in many respects retarded rather than aided the PWA's institutional development, complicating the formation of stable decision-making structures, and complicating the task of producing stable knowledge. This has even to some degree been true of those plans funded with the explicit purpose of integrating previous ones.[42] The problems here, it should be emphasised, have not arisen due to any deficit of information, but because of its social and political organisation.

Quite apart from the problems of donor-related issues of duplication and contradiction, it is highly questionable whether medium- and long-term planning is feasible or appropriate in the present Palestinian context. For anything other than short-term planning, the PA insists on working with the assumption of full Palestinian sovereignty over the whole of the West Bank and Gaza; it insists on assuming control of all natural resources; and it also insists on assuming that all Israeli settlements will come under Palestinian control, and be inhabited by Palestinians. These understandable political demands constitute insuperable obstacles to the task of producing institutionally stable and useful plans. Given the uncertainty of the interim period, and given also the PWA's dependence on donors and contractors, it is hard to see at present how it would be possible to construct stable administrative structures, or to produce medium- or long-term plans and policies that are of anything other than purely hypothetical value.

GOVERNING SYSTEMS AND SUPPLIES

If the PWA has had far from unproblematic relations with donors, thereby complicating the tasks of planning and institution-building, the West Bank Water Department has encountered even greater difficulties in its relations with municipalities, village councils and even individuals, and in its work of governing local water systems and supplies. For their part, municipalities have often competed with one another over the control of resources and supplies, and at the same time been unable to control the conduct of errant individuals. Here we consider some examples of these internal Palestinian water sector dynamics.

As already noted, the Bethlehem water authority is under the close control of the PWA. Hebron municipality is a different matter altogether, however: PWA officials, and international contractors with experience of working with it, often ruefully speak of it as a 'state within a state'.[43] There is fierce conflict between Hebron and some surrounding municipalities and village councils over the control of local resources, key amongst them water. For instance, Hebron municipality has long owned and controlled two wells to the south of the city, these supplying both Hebron and the town of Dura, 10km to the west. Dura is wholly dependent on these Hebron-controlled wells for its piped supplies. However, in the absence of national pricing controls, Hebron is free to charge the water rates that it can get away with, and is also free to cut water supplies to Dura and use them for its own internal purposes, as and when it sees fit. In consequence, two-thirds of Dura's water supplies are met by private water tanker companies, who charge much inflated prices of $2-4 per cubic metre.[44] By no coincidence, Dura has one of the lowest per capita water supplies in the southern West Bank.

Of course, most water sources in the southern West Bank are not controlled by municipalities, but by Mekorot. These waters are bought from Mekorot by the West Bank Water Department, which then supplies municipalities and village councils, who in turn are billed by the Water Department for water received. However, municipalities and councils are loath to pay for their water, such that between 1995 and 1998, for instance, the Water Department amassed NIS 30 million ($8 million) worth of debts from non-payment for water supplies.[45] By 2002, total Water Department debts were NIS 110 million ($24 million).[46] Municipalities and village councils typically claim that they are unable to pay the Water Department, since the extent of individual non-payment within their own municipalities is so high. Water Department and also PWA officials contend, however, that quite aside from the problem of individual non-payment, local authorities are hoarding payments, and diverting these funds into road, electricity and other local projects (and no doubt also into personal bank accounts).[47] Unwilling to cut water supplies to municipalities, and in the absence of a national authority that is willing to enforce the rule of law, the Water Department is powerless to prevent non-payment. Indicatively, it is Hebron municipality that is the Water Department's largest debtor, owing, as of April 1998, NIS 13 million ($3.5 million).[48]

To illustrate some of these inter-municipal and municipal-national tensions in more depth, consider an incident that occurred in the village of Duwarra during July 1998.[49] This village, as already noted, is fortuitously located alongside the 16-inch transmission line leading from Herodian to Kiryat Arba and Hebron, and receives a constant supply from this pipeline, even during the height of summer (see Figure 3.3). Hebron, on the other hand, lies at the very end of the line, and thus receives from it only that which is not used first either by Palestinians living alongside it, or by the settlers of Kiryat Arba. Each summer, Hebron finds itself suffering prolonged shortages, and often seeks to take matters into its own hands as a result. As for Duwarra, it not only has constant access to a major distribution line, but making matters still worse, during summer 1998 several of its residents had constructed illegal connections to the main line, which they were using to steal water for irrigation. This broke a longstanding agreement with the West Bank Water Department, which ten years earlier had given Duwarra a legal connection to the village on condition that all illegal activity would cease. This agreement had now been broken, and hence Duwarra aroused the enmity of both Hebron municipality and the Water Department. Thus on 30 July 1998, a group of officials from the Water Department and Hebron municipality arrived in Duwarra, intent on closing the supply valve – the legal connection to the 16-inch main line – located in the centre of the village; they were accompanied by officials from Kiryat Arba, not averse to entering into limited co-operation with local Palestinians (there is indeed a well-established pattern of co-ordination between Hebron municipality and various Israeli authorities, as we shall see at several further points during this chapter).[50] This group of officials was immediately repelled by villagers, and prevented from turning off the valve; they left, to return several hours later accompanied by a 16-strong Israeli–Palestinian patrol, and an additional 20 Palestinian police officers. By then, a third of Duwarra's 1500 populace were waiting for them, ready to defend their water valve. A stone-throwing battle ensued, and several Palestinian police consequently ended up in hospital with broken limbs (as is usual in such cases, the Israeli officers stood back, and let the Palestinian officers take responsibility for policing their own population). Order was restored only with the arrival of members of the Palestinian security agencies – not just one of them, but three – whereupon it was decided that the village valve would remain open, and that Duwarra's residents should be allowed to keep their water supply.

Such incidents were common during July and August 1998, and are typical of the southern West Bank as a whole.[51] This particular incident, though, highlights several prevailing tendencies. It illustrates, firstly, the sheer complexity of local inter-municipal water conflicts, typically involving *ad hoc* alliances between national Palestinian authorities, police forces, security agencies and even Israeli settlers. As in this case, the extra-judicial security agencies are often the ultimate arbiters, and hence outcomes tend to depend on who can recruit security officers to their cause. By no coincidence, several of the leading figures within Duwarra's village council were members of local security agencies.[52] Outcomes are also strongly affected by the extent of physical control exercised by the various parties. In this case, Duwarra's residents were empowered not only by their proximity to a main supply line, but additionally by the location of the supply valve in the centre of the village, which makes supplies all the easier to defend (the supply valve to the neighbouring village of Eddaiseh is, by contrast, located well outside the village; this valve was closed a number of times during July and August 1998, each time without the knowledge of its residents). In these conflicts, the Water Department is very often powerless. It lacks the means to enforce its regulations and policies, and generally has little option but to form unpredictable *ad hoc* alliances, and to engage in local politics, just like the local municipalities. Far from being a regulatory body standing above society, the Water Department here is one protagonist amongst other 'strongmen' vying for power and control. It is not clear in this particular case whether Duwarra was given official notification that its supply was to be cut, but if there was such notification, it evidently carried little weight.[53]

Throughout the southern West Bank, Palestinians and also Israeli settlers continually steal water from main transmission lines, and from internal community networks. In and around Duwarra, villagers often cut into main lines, their connections being hidden underground, or buried under piles of rubbish. Others take the easier route of disconnecting their water meters. Some take their water meter out by night, spend several hours irrigating their crops, and then reconnect it before morning. In somewhere like the Dheisheh refugee camp, to be discussed at length in the next chapter, there are hundreds of underground illegal connections.[54] Prior to 1998 in Dura, the town's main private water salesman was routinely filling his water tankers with supplies stolen from the pipeline feeding Dura from Hebron, and receiving protection in this, it seems, both from Hebron municipality and the Israeli authorities.[55] Even when their water use is metered, residents often refuse to pay for it, typically claiming that they are unable to do so. After years of resistance against the Israeli occupation, during which time the PLO often advocated non-payment of water bills, people are loath to change their habits. Given this deeply embedded ethic of resistance within Palestinian society, national and municipal water authorities, and village councils too, all face problems in controlling and disciplining their customers.

That this is the case can be clearly seen from the Bethlehem municipal area where, during 1998 and 1999, the WSSA and their French contractors attempted to implement a 'yield improvement' project by clamping down on water theft, enforcing payments, and improving the operating efficiency of the internal network.[56] Central to this endeavour was the installation of 10,000 new and more efficient household water meters, replacing meters that were as much as 30 years old, and cutting under-registered water use by as much as 35 per cent.[57] In replacing these, the WSSA's (and PWA's) aims were to increase revenues, to pave the way for the development of a profit-making institutional system, and to produce docile consumers who would use water only in accordance with their willingness to pay. Yet the problems here were twofold. To start with, the very fact that these new French meters were more efficient necessarily meant that water charges increased enormously. Moreover, these new meters, while very good at metering water, also proved highly adept at measuring air, with the consequence that, during the summers of 1998 and 1999, people who had not seen piped water for months found themselves in receipt of extortionate water bills. Given that there was often little water in the pipes, and given also the poor design of Bethlehem's internal network (especially its dearth of air release valves), there was little the WSSA could do to prevent such over-metering – all water meters will read air passing through them, just some more accurately than others! The result was predictable. Lacking trust in the water authority, and often unwilling to pay higher rates for still irregular water supplies, many of the new water meters were damaged or disconnected. In an attempt to prevent interference, the authority responded by installing steel boxes around their meters; and also, during 1999, began cutting off supplies to those guilty of non-payment.[58] Whether the WSSA will at some point succeed in disciplining its water users is an open question. Yet what is clear is that the WSSA's attempts to do so were being hindered both by the overall dearth of water supplies received (since this necessarily increases the volume of air within the pipes, and also inspires reticence to pay high rates for water supplies) and by a lack of trust in a politicised and faction-dominated institution (on the streets of Bethlehem, people often voice suspicions as to why particular rock-cutting factories and hotels receive constant supplies of water, and even suggest that the WSSA has a stake in the private water-tanker trade). As in Hebron and Duwarra, the neo-patrimonial character of contemporary Palestinian politics and society, and the resistant activity of non-institutional and even individual actors, render the tasks of governing water systems and scarce supplies incredibly difficult.

CONSTRUCTING INFRASTRUCTURES

Combining the emphases of these two previous sections, we can now consider the problems faced by the PWA in constructing its new water supply infrastructures. Irrespective of my earlier comments about the feasibility or otherwise of institution-building and planning under Oslo, the majority of donor funds have been used for the purposes of infrastructure-building and supply development. During the period 1995–98, for example, 80 different donor projects were dedicated to the improvement of supply in the West Bank and Gaza.[59] Of these, the largest was and remains USAID's Water Resources Program in the West Bank, which has concentrated primarily on implementing the terms of the Oslo II water accords by developing new and additional resources from the Eastern Aquifer, primarily for the Palestinian communities of the southern West Bank. The first phase of this large-scale project was due to be completed by June 1999, providing an additional 7.2 mcmy of water to the Hebron and Bethlehem areas.[60] Given that total water use in these areas was only 10.6 mcmy in 1996, this was to represent a monumental increase in water supply levels.[61] Unfortunately, however, at no point has this full additional supply ever been received.[62] Explaining this requires that we consider, once more, the issue of the institutional power of the PA, both in relation to international donors and contractors, and in relation to errant municipalities.

Phase One of the Water Resources Program in the southern West Bank involved the construction of four deep wells into the Eastern

Aquifer, in the Herodian area to the south of Bethlehem (known as PWA1, PWA3, PWA11 and Hundaza; see Figure 5.2); the construction of 32 km worth of transmission line, leading from these wells to Bethlehem and Hebron; and the construction of two large storage reservoirs and a booster station. The scale of this work would, in Palestinian terms, be enormous: the supply line would mostly be of 36-inch diameter (before this, the largest pipelines in the southern West Bank were of 16-inch diameter, feeding settlements such as Kiryat Arba; those pipes directly supplying Palestinian communities were at most 12-inch diameter); and one of the reservoirs, located near Halhoul, would have a storage capacity of 25,000 cubic metres (previously the largest being of 3875 cubic metres capacity).[63] The system as a whole would have a capacity of 81 mcmy.[64] Moreover, this new system would, once constructed, become the property of the PWA. It would also be the first system 'to serve the Palestinian population solely', and would thus be controlled, in accordance with the Oslo II terms, solely by the Palestinian Authority.[65] The whole project would cost $72 million, this sum having been given by USAID in the form of a grant.[66]

Work on the project began in the West Bank in July 1996, the contract for overall management of the project having been awarded to a consortium of mainly American firms, collectively known as CDM/Morganti.[67] Detailed designs for wells, pipelines and so on were completed during 1996, and JWC permission for the proposed facility developments was granted, with some changes, by September 1997; thereafter, construction sub-contracts were awarded, once again to US firms.[68] Construction of wells and pipelines began in January 1998.[69] The project had been delayed in its early stages because of Israeli objections to several of the proposed well sites; nevertheless when construction began in January 1998, it was still expected that the project would be completed on time, by June 1999. That it was not was a result of three distinct, though related, sets of problems.

The PWA, first of all, had trouble regulating the actions of the international and Palestinian contractors working on the project. The American corporation ABB-SUSA, which had been sub-contracted by CDM to construct the transmission lines, reservoirs and the booster station, in turn subcontracted most of their work to Palestinians, who were supervised in their work by just three international ABB-SUSA engineers.[70] As a result, there was insufficient supervision of the construction work, with neither ABB-SUSA, nor CDM, nor the PWA keeping a regular eye on the subcontractors.[71] This led, wittingly or otherwise, to some of the design specifications being violated.[72]

Insufficient air-release valves were placed along the main pipeline, thus increasing the likelihood of vacuums developing within the pipeline, and of the pipeline imploding as a result (this did in fact happen during summer 1999).[73] There was insufficient testing of the pipeline while it was being constructed.[74] In addition, much of the engineering work was done shoddily, such that both the main line and one of the reservoirs leaked badly.[75] The long-term effect is that the PWA's first autonomous water supply system is in various regards structurally deficient, and will also, in consequence, be very difficult to operate.[76] Just as seriously, faulty equipment was installed in the well pumps such that, within around a year of them coming into operation, pumps in three of the four new wells had broken down. As of summer 2002, all three wells were out of operation (the main contractors, CDM, have accepted responsibility for this, with some responsibility also having been accepted by the US subcontractors Techmaster for fitting the wells with sub-quality equipment and violating design specifications).[77] Put simply, the reason for this is that while contractors and subcontractors are motivated above all by profit – it was because of this, after all, that the engineering work was subcontracted from American to Palestinian firms, and it was for this reason that short-cuts were taken, and that supervision was kept to a minimum – the PWA lacked the institutional means to regulate and control these quite normal private sector activities.

The project also encountered 'numerous landowner difficulties'.[78] In constructing their wells, pipelines, reservoirs and so on, the PWA and contractors necessarily had in some places to buy land, and in others to dispense compensation for damage caused. Landowners often strongly objected to the proposed constructions, and especially to the seizure of land and the uprooting of trees. Throughout the West Bank, trees are of great symbolic importance, and their uprooting recalls and resonates with Israeli occupation policy. Moreover, in the southern West Bank, trees are of material-practical significance, fruit and olive trees being one of the mainstays of the local agricultural economy. On several occasions during the implementation of this project, farmers sought to physically defend their land against contractors. The problems here were exacerbated by the fact that most construction areas were in Area C, under full Israeli control, and were thus out of bounds to Palestinian police; hence construction workers had no police escort, and the Palestinian police could do very little to prevent civil resistance. Moreover, these landowner problems were also exacerbated by the fact that contractors had something of an interest in their work being disrupted, in that they would then have to be paid for lost days, and for additional days' work.

On one occasion, PWA officials witnessed contractors attempting to persuade a landowner that the PA was not going to pay compensation for land seized, their aim being to provoke him into disrupting construction work; PWA officials believe that this was not an isolated case. In summary, the PWA has had recurrent problems in controlling non-compliant landowners, partly because agricultural land and trees are so important within the Palestinian imaginary and the southern West Bank economy, and because civil resistance is so much a part of everyday Palestinian life; but also because the PWA is institutionally weak in relation to private sector contractors, especially in the Israeli-controlled Area C.

Besides these landowner and contractor problems, a third problem lay in a local political conflict that developed between the municipalities of Hebron and Sayyir, and between them and the PA.[79] To explain this it needs noting that, just before the signing of the Oslo II Accord in 1995, Hebron municipality had reached agreement with the German state development bank, KfW, and had also received permission from the Israeli Civil Administration, for the DM 11 million construction of two wells and a 16-inch pipeline feeding from them to Hebron city.[80] These two wells (known as Hebron 1 and 2; see Figure 5.2) were to be located in the Herodian wellfield, not far from the planned USAID-funded wells. The 16-inch pipeline from them would pass through the town of Sayyir on its way to Hebron, as also would USAID's 36-inch line. Unlike the USAID-funded systems, however, these KfW-funded facilities would be owned and controlled by Hebron municipality.

KfW's two wells were drilled during 1995 and 1996, a couple of years earlier than those drilled by USAID. Nevertheless, the PWA understandably wanted co-ordination between the two projects, favouring the construction of a single pipeline from Herodian to Hebron that would convey water both from the two Hebron municipality wells, and from those being drilled for the PWA. Hebron and KfW declined, however, partly because their project was a couple of years in advance of USAID's, but also for more obviously political reasons. KfW sought, on the one hand, to 'fly the flag' over its project, as is seemingly typical of donor agencies. Hebron sought, on the other hand, to maintain sole ownership and control of this new infrastructure, both because this would enable it to secure its own water supplies, and because it would give it greater political leverage in relation to surrounding municipalities and the PA. However, construction work on the KfW pipeline was soon delayed by the residents of Sayyir, who harboured grievances against Hebron, and objected to the fact that Hebron alone would be receiving new and additional water from the project. All that Sayyir was to receive from the project was the digging up of one of its few paved roads. Thus Sayyir residents physically prevented construction work, and the project remained at a standstill throughout 1997. Matters got still worse in 1998, following the appearance of 36-inch pipeline segments along the very same stretch of road where KfW were still planning to lay their 16-inch pipeline. The Sayyir road was now seemingly to be dug up twice, with pipes to be laid on either side of the route to Hebron. Several stone-throwing incidents ensued, as did complex and protracted negotiations between the municipalities of Hebron and Sayyir, the PWA, the Ministry of Local Government, and USAID, KfW and their respective contractors. Eventually the conflict was resolved such that the two pipelines would in certain stretches be laid within a single trench. This has since been done, with the result that, during summer 1999, one of the two KfW wells began supplying water to Hebron. However, both the KfW and USAID projects were significantly delayed as a result of the conflicts between the two municipalities. Although the conflict was eventually resolved, Hebron managed to maintain ownership and control of its infrastructures – clearly testifying to the weakness of the PA in relation to municipalities, to the non-institutionalised dynamics which typically characterise the Palestinian arena, and to the ways in which international donors and contractors can complicate processes of infrastructure-building and state formation.

It was not until autumn 1999 that Phase One of USAID's work in the southern West Bank began yielding any new water. Until this point, the only new supplies to have been received since the signing of the Oslo II Accord in 1995 were an additional 1 mcmy provided by Israel (as stipulated in the agreement), and the water flowing from the new Hebron well in the Herodian area (which in any case had been planned and agreed before Oslo II).[81] Thereafter, however, some new water did come on stream. The water situation in Bethlehem improved markedly in 2000, partly because of the new supplies, and partly because of the already-mentioned yield improvement project.[82] In surrounding villages, however, supplies remained erratic, as they did to an even greater extent in Hebron and vicinity. Of the two Hebron wells which had been the source of so much controversy, one turned out to be dry, and the other was fitted with an inappropriate pump, and was soon out of order.[83] Conflicts remained as severe as ever between Hebron and other local municipalities.

PWA officials insist that the problems encountered in the southern West Bank during implementation of Phase One of the USAID project need not recur, since they resulted primarily from the specific nature of the contracts that had been agreed with USAID, and from the fact that the PWA was still in an early stage of institutional development.[84] This may to some degree be so: the PWA will hopefully have learned from these experiences, such that Phase Two of the project runs more smoothly. Nonetheless, given the structural contexts within which the PWA has to operate, there are clear limits as to how much could ever be achieved through institutional learning. The PWA operates within a deeply fractured social and political system, is almost wholly dependent on international donor funds, and is limited by (and also a product of) a political process that has, all along, prioritised the development of a repressive security apparatus, rather than strong administrative institutions and the rule of law. Given all this, there is little reason to think that an Oslo-constrained PWA was ever likely to overcome its regulatory and institutional troubles.

Yet if that was so of the Oslo period, it has become increasingly the case since September 2000, and in light of the escalating violence, economic devastation and humanitarian emergency situation across the West Bank and Gaza. Non-payment levels have massively increased, both by individuals and municipalities. Municipalities warn that, faced with endemic non-payment and therefore forced to make cut-backs, many of their new and rehabilitated infrastructures will be out of operation within a couple of years.[85] A World Bank-funded utility-building project – developed with the specific aim of bringing resistant municipalities to heel, and involving in the southern West Bank the creation of a single water utility for the entire region – has proved impossible to implement, in large part because international contractors have repeatedly claimed *force majeure*, and have been unwilling to travel around the West Bank.[86] Intent on prioritising emergency needs, relief organisations such as Oxfam and the Red Cross are now largely bypassing the PWA, opting instead to deal directly with municipalities and village councils, and are putting their money into water-tankers rather than pipelines. All of this is readily understandable in light of the desperate situation in the Territories since September 2000. The inevitable result, though, is a further weakening of the central Palestinian water institutions, a further fracturing of the Palestinian water sector – and with it a further waning of the chances of central Palestinian institutions ever being strong enough to resolve the Palestinian water crisis.

CHAPTER 8

The Arts of Getting By

But we should not end on such a wholly discordant note – for while the central and even municipal institutions of the Palestinian proto-state are clearly very weak in their capacity to manage and regulate their water sector, the same cannot be said of ordinary water users. Where local and national institutions fail in their tasks of providing regular and predictable supplies, water users have no option but to engage in alternative forms of supply and demand management. And what we find, when we consider these in detail, is that ordinary Palestinians are highly adept at 'getting by' in the face of water shortages.

In most expert discourse, the Palestinians are represented as facing a 'water crisis', as being on the verge of a precipice that they – and sometimes the Middle East as a whole – are about to fall over.[1] I, too, have used the phrase 'water crisis' repeatedly in this study. Nonetheless, crisis representations do tend to elide, amongst other things, the fact that the work of coping with water shortage is part and parcel of Palestinian West Bank life. Whereas in expert discourse (and also admittedly in most Palestinian nationalist discourse) the Palestinian water situation is depicted as extraordinary and indeed catastrophic, water supply cuts and the coping practices that they necessitate are in another sense quite mundane. Water experts tend to ignore these aspects, relying instead on aggregated quantitative indices of water shortage. Water, in most of the expert literature, is simply consumed by abstract domestic, municipal, industrial and agricultural sectors, such that the diverse practices of drinking, cleaning and irrigating, let alone those of fetching and storing, simply get subsumed and hidden under the technical heading of 'water demand'. Neither expert representations nor the term 'crisis', however, in any way manage to portray the real, day-to-day experiences of those people who suffer them, or indeed to capture the chronically grim normalcy of water shortages.

On those few occasions where 'ordinary' water users do figure in expert accounts of Middle Eastern water problems, it is usually in pejorative fashion. The work of Tony Allan is typical in this regard. Water, he observes, is of immense 'salience ... in Middle Eastern cultures', this being readily apparent from the 'books of the major religions', especially those of Islam.[2] Perceiving water to be a gift from God, Muslims believe 'that water should be free or at least very low cost', and assume that people have certain 'rights ... to low cost water', and even an 'entitlement ... to free water'.[3] The pertinence of this, says Allan, is that such beliefs and perceptions can often stand in the way of modern water management practices. Because of their traditional ideas, Middle Eastern communities experiencing transitions into water deficit are simply 'not equipped to deal with the new circumstances' they face.[4] There is often 'fierce resistance', above all, 'to the idea of water being an economic resource'.[5] Moreover, 'such is the strength of the existing beliefs that they easily withstand the assaults of new knowledge brought by outsider professionals and scientists'.[6] Ordinary water users and their traditional culture are, by this account, a straightforward impediment to rational water management.

It barely needs saying that these claims, which are typical of so much culturalist reasoning about Middle Eastern politics and society, involve very simplistic generalisations about the region's peoples; are inattentive to the heterogeneity and local variability of attitudes and beliefs; ignore the fact that traditions are constantly being reinvented, as well as the question of how traditions are reproduced and transmitted (if indeed they are); and disregard the practical, material dimensions of Middle Eastern culture, depicting it purely in ideational terms. It is no doubt the case that the Qur'an has plenty to say about water, and that Islamic texts can be utilised in formulating and also marketing water use principles and water laws.[7] However, it is another thing altogether to argue that Muslim 'peoples are imbued with the precepts of Islam as articulated in the holy Qur'an,' and that they are thus bound by traditional beliefs about water.[8] Such a claim not only ignores the differences in attitudes between, say, urban middle-class and bedouin Muslims. It also makes the mistake of understanding beliefs in over-structuralist terms – eliding the fact that beliefs are often inconsistent, and understating the flexible agency of social actors. As we shall see, social actors are highly responsive and adaptive to contexts, such that culture and tradition are not the barriers to rational water management that Allan and so many other water experts typically present them as being.

With the aim of arguing thus, the main body of this brief final chapter presents an ethnographically informed account of everyday Palestinian water use practices in the southern West Bank. I focus almost entirely on practices relating to domestic water supply, saying very little about those relating to agriculture and irrigation.[9] We first consider a series of snapshots of the micro-scale material reality of water shortage, concentrating in particular on the situations in Dheisheh refugee camp, Bethlehem, and in the village of Quasiba to the south. Having let these snapshots to some extent speak for themselves – as if this were at all possible! – only then do we consider their practical and conceptual implications.

COPING WITH CRISIS

We start in Dheisheh refugee camp.[10] Located on steep western-facing slopes on the southern outskirts of Bethlehem, Dheisheh was, from late 1995 onwards, in an area of full Palestinian Authority control (Area A). The camp is crowded with *ad hoc* structures, with dwellings which were first built during the late 1950s by the UN Relief and Works Agency for Palestinian Refugees (UNRWA), but which have since evolved and been extended in all manner of ways. Between the chaos of houses run sharp, narrow streets and winding alleys; above the houses one finds a strange *mêlée* of water tanks and satellite dishes. Just under 10,000 people live in the camp, all refugees from, or descendants of refugees from, the 1948–49 war with Israel, all of them tracing their roots back to villages in the foothills of Jerusalem.[11]

Every summer during recent years until 2000, Dheisheh had been plagued by severe water shortages, and this despite the whole camp being connected to Bethlehem's supply network. Throughout the summers of 1998 and 1999, Dheisheh would receive water for four to five days, and then go around 12 days without. There was nothing routine or predictable about this, however. During late July and early August 1998, the camp went 15 days without piped supplies, and even when the water came back on, it lasted for only three days. No-one in the camp knew precisely when the taps would be turned on and off. Moreover, even when water was being supplied, one could never be certain about the water pressure. Water was generally received for five days every 17 at the foot of the camp, adjacent to the main road linking

Bethlehem with Hebron, but further up the camp supplies were much less frequent. The higher up in the camp one went, the worse the supply conditions would become. One house or street would receive water for three days every supply cycle; the next house or street up, for just one day in 20. During the summers of 1998 and 1999, the highest areas of the camp, on the ridge looking south-east over the valley of Artas, went without piped supplies from mid-May onwards.

Irrespective of its location in the camp, however, every household would face some degree of water shortage. With each supply cut, families would be left to wonder whether they could last out. Some would opt to buy tankers full of water from local merchants, using them to fill the cubic metre steel tanks on their roofs. For most, though, such water would simply cost too much – anything up to NIS 200 ($50) for a 10 cubic metre delivery – and in any case, water merchants would not be at all keen to make deliveries within the camp and would often refuse to do so, claiming that the streets were too narrow. Thus most camp residents would resort to other means. People living higher up the camp would rely, in the first instance, on their fellow residents' willingness to give and share water. Young children would scour the camp with cooking pans and plastic bottles, asking for water. Women would carry bucket-loads up the camp's steep alleyways to their homes. Clothes would get taken from house to house to be washed. Families would go to the houses of friends and relatives to wash and take showers. With the situation deteriorating, and more and more of the camp running dry, people would start looking further afield. Some would take their cars to the local gas station, in the hope of getting water for free; others would go to Aida refugee camp on the far side of town, where water was provided from an UNRWA standpipe; still others would get water from, or do their cleaning and showering at the homes of friends and relatives living in and around Bethlehem; and many would get their water from one of the many local springs, such as that in the village of Artas.

Some people would make their way to the spring, Ayn Artas, by walking up and over the camp, then down the other side into the valley and village. On summer evenings during 1998, women and teenage girls would trudge back and forth between the camp and the spring, buckets loaded and balanced on their heads. Others, often whole families, would journey there by car, heading south out of the camp along the dusty road through Al-Khader, and then down to the village past the ancient ruins of Soloman's Pools. Some, usually older, men would ride to the spring and carry their water by donkey. Still others, mainly children, would push supermarket trolleys filled with water bottles, or load up ingenious home-made contraptions and drag them along the floor.

The spring itself issues forth into a deep, four-walled concrete structure, whereupon its waters flow into and through a shallow, open channel. On a typical summer evening during 1998, the *ayn* was flooded with people in search of water. Hordes of men, women and children would jostle around the spot where the clean spring water emerges, each with their own personal range of buckets, bottles and canisters. Amidst the physical struggle, some would fail to fill their containers, and give up and head elsewhere. But there would constantly be more people arriving. Further away from the source, down the channel, women would wash clothes and talk, while young boys would jump and splash around, pouring water over themselves and one another, laughing and screaming. The spring would bear evidence of hardship, but for some at least it would also be a site of resilience and playfulness.

Back in the camp, water would be used with the utmost care. For those at the foot and in the middle of the camp, daily life would wait for the piped waters to return, and would bear the ever-changing marks of the supply cycle. Piles of dirty laundry would grow and grow. Grey water would be reused for the toilets, and for garden plants and vegetables. In many houses, cooked meals would be sacrificed in order to avoid a pile-up of unwashed dishes. People would buy more bottled water, cola and other soft drinks. Often children would be forbidden from playing outside in the streets, in the hope that they would avoid getting dirty. For the lucky families who still had piped or tanked supplies, taps would be fitted with pieces of sponge, such that barely a drop would be lost. Women, in particular, would wait for the water to return, for the chance to do their washing and cleaning. Children, but adults too, would listen eagerly and carefully for the distant gurgle of water down the pipes. The supply cycle would make little odds to those living higher up the camp, however; they would go without piped supplies for the duration of the summer.

With the final arrival of water through the pipes, there would develop a quiet frenzy of joyous, earnest activity. The lower parts of the camp would suddenly be bedecked with metre after metre of rubber tubing. These areas would vibrate to the solid drumming of motors, as people set about the business of pumping water up into the empty steel vats on their roofs. Sometimes the pressure would be too low, the water would come out in splutters and starts, and the whole day would be spent plugging and unplugging motors. At other times, the work would involve little more than clambering around, transferring tubes from one tank to another. With everyone filling up their tanks, the camp would literally devour water, with the result that those areas beyond the foot of the camp would still remain without. People living slightly higher up would wait, half-anxiously, for those below to fill their tanks, and for the water to reach their own houses, always unsure of whether the pressure would be sufficient for the water to reach them. Some of them, in the meantime, would lay lengths of rubber tubing along the streets, and endeavour to achieve by themselves what the mains system was failing to do, namely, to pump water up to their own homes. Scores of men, and even more so women and children, would come to the foot of the camp, where in a vibrant festival of water, they would wash and clean and shout and fill their buckets. Water would be used with relative abandon, an abandon born of day after day of shortage, and out of the clear knowledge that the water would soon once more be cut, and that one had better make the most of it while it remained.

Each family and household would face its own unique range of problems. Water supply, as we have already seen, would vary from one house and one street to the next; but the complexities do not end there. Just as supply would vary, so too would demand for those supplies, from those houses which supported a dozen children, to those inhabited by a single individual; from those that made use of flush toilets and automatic washers, to those that did not. Difficulties would vary, moreover, not just between locations, between households, but also from one moment and one year to the next. Who could know when and where a new checkpoint would be established, when a closure would be imposed, or when a demonstration would make the route to Aida or to Artas impossible? Who could guess for how long the pipes would remain dry? Who dared think how much this year's water merchants would charge? And, above all, who could be sure whether the winter rains would be sufficient to replenish the local springs? Certainly not the residents of Dheisheh, who found to their dismay during the summer of 1999, that Ayn Artas had dried up, and that new and more distant sources of water would have to be found. Each summer Dheisheh's residents would face the constant and general threat of summer water shortage, as well as sets of problems that were ever-changing, locally variable and highly particular.

Residents would at times complain that 'Bethlehem has water, Doha has water, Beit Jala has water all the time; it's a special problem just in Dheisheh camp'.[12] But this was not the case. Like Dheisheh, most of the rest of Bethlehem would receive water for less than one week every three over the summer. For instance, the town of Beit Jala which, like Dheisheh climbs steeply up a mountainside, was marked during 1998 and 1999 by supply conditions that varied from one street to the next.[13] At its foot, water was generally received every third week, while towards the top, most households would end up going months without piped supplies. Others, higher still, were fortunate enough to share a network with the local District Co-ordination Office (where, prior to the new intifada, Israeli soldiers and Palestinian police would co-ordinate their patrols), and would hence receive near-constant piped supplies. Water problems were less severe in Beit Jala than in Dheisheh, but this was not due to any greater regularity of supplies. Rather it was because Beit Jala residents are generally much wealthier than those of Dheisheh, and thus are much more able when necessary to buy tanker water supplies; and because most houses in Beit Jala are constructed above vast underground cisterns with capacities of 30 cubic metres or more, such that the people living in them can much more readily cope with lengthy supply cuts (thus also, while the houses of Dheisheh are typically festooned with as many as a dozen water tanks – excluding, that is, those houses that are too weak and unstructured to withstand the weight of more than a couple – those of Beit Jala mostly have only a few). While Beit Jala's water problems were not as severe as those of Dheisheh, this had less to do with the quality of supply than with the generally greater capacity of Beit Jala residents to adapt to and cope with water shortage.

Yet if prior to 2000, Bethlehem was afflicted by water shortages, the situation was typically much worse in the rural communities of the southern West Bank. Take, for example, the case of the isolated village of Quasiba, already touched on in chapter 3.[14] Throughout the 1990s and still to this day, Quasiba receives its water via the town of Sayyir. Usually in the summer Sayyir receives a relatively good supply of water – typically five days on, then five days off. Quasiba, by contrast, receives nothing. By mid-August 1998, for example, the 2-inch tube running along the roadside and over the hill between Sayyir and Quasiba had not seen water since April. Every year this dearth of water causes enormous difficulties. At first people would rely on water collected from the winter rains, on water that falls on their rooftops and is then channelled and stored in underground cisterns, but this would never provide sufficient for long. Some of the poorer villagers would resort to using and even drinking water from the old, barely maintained spring in the centre of the village, notwithstanding the fact that it was recognised as polluted and maybe dangerous. Those with cars, and with good family or factional or security service connections, would collect water from acquaintances in

Sayyir, others from further afield. Still others would have their own personal mini-tankers, which they would convey by tractor to Ayn Tequa, a spring which is at the same time a recognised filling station for private water-tanker businesses. Demand for the spring's waters would be incredibly high, however, and hence people would have to wait for five, six or seven hours to fill their tankers.

Given the extremity of this situation, it is hardly surprising that some people would take matters much more fully into their own hands. One such is Ziad Tarawa, in the uppermost house on the eastern side of the village. Twenty metres behind Ziad's house one finds an enormous cavern, a motor perched on its crest. Although it looks natural, it is in fact Ziad's creation, having been excavated by him and others, with the aid of drills and explosives, in 1994. The cavern, 9 m deep and as much as 12 m wide and long, is filled, every winter, by water flowing through the mountain rising sharply behind. Rubber tubing leads from the cavern to rooftop tanks on Ziad's and five other houses, the cavern's waters being sufficient to meet the total domestic water needs of these six households. Such are the lengths to which rural West Bankers often have to go in order to secure their domestic water needs.

Beyond the above, we have in previous chapters already encountered a range of other water use and management practices: the resort to force to maintain physical control of water facilities and supplies (as in Duwarra), and also the theft of water either for the purposes of meeting domestic needs (as in Dheisheh), of filling privately owned water-tankers (as in Dura), or of irrigating fields (as once more in Duwarra). From the licit to the illicit, what by way of summary can be said about these diverse practices of water use?

EVERYDAY AGENCY

A first obvious point to make is that the southern West Bank is home to a kaleidoscopic variety of water-related patterns and practices. Above all, practices vary in response to supply conditions. They thus vary not simply between rural and urban areas, but from one village to the next, from one town to the next, and even from one street to the next. Practices also vary, however, in relation to people's differential modes of access to alternative supplies, and in relation to levels of household demand (which even within Dheisheh refugee camp vary wildly, from those houses with automatic washing machines, wet toilets and showers, to those without). Practices do not simply vary between people, households and communities, though, but also across time. They differ in response to the nature and phase of the supply cycle, and they vary from season to season, and from year to year. They display an immense variability.

Secondly, Palestinian water use practices in the southern West Bank are far from traditional. One of the more startling things that one comes across in the West Bank is the frequent juxtaposition of the old and the new. Satellite dishes and water tanks sit side-by-side on rooftops; water is collected from Ayn Artas by car and donkey, by bottle and supermarket trolley. The sight of women with buckets on their heads often carries with it resonances, for the European observer, of a traditional society, and indeed there may well be some vestiges of this. According to one Palestinian water expert, the fetching of spring water still provides opportunities in some rural communities for women to escape from the clutches of the traditional, male-dominated Palestinian society, 'the women refuse to have water in their houses in some villages, because the social structure doesn't allow them to leave the home,' he asserts; 'the women's only opportunity to leave the home is to go to the spring… it's like a club to the women'.[15] Be that as it may, in places such as Dheisheh the women carrying water are often educated professionals, employed as schoolteachers, or working with NGOs or the Palestinian Authority. It is simply not the case that Palestinian water use practices in places such as Dheisheh, Duwarra or Quasiba are mired in tradition, or are structured by age-old patterns of culture.

Julie Trottier argues that irrigation practices in rural West Bank communities are primarily structured around longstanding local customary laws that vary from village to village.[16] This is readily conceivable in some cases. However, it is certainly not the case in villages like Duwarra, where water use patterns and practices are principally structured by the existence of a major pipeline; and nor is it the case in the West Bank's towns and refugee camps, where the majority of the Palestinians live. For the most part, such 'traditional' forms have been swept aside as a result of dispossession, occupation, economic modernisation and the destruction of Palestinian agriculture, or at the very least have been adapted to and transformed by these interactions.

Thirdly, the water use practices considered here are marked by an enormous degree of flexibility and adaptability. Against those accounts that depict culture as a barrier and impediment to rational management, or that suggest that Middle Eastern communities are 'not equipped to deal with the new circumstances' that they face, the practices considered here are marked by a high degree of eminently rational and context-sensitive agency.[17] In the southern West Bank, after all, it is individuals, households and communities – rather than expert institutions – that in the final analysis manage to govern and administer water supplies in accordance with local social needs. Where municipal and national institutions fail in the task of providing regular and predictable supplies through their large-scale technological networks, water users engage in an alternative form of supply management using their own micro-scale techniques: searching for water, stealing it if necessary, collecting it in bottles and canisters, conveying it by hand and car, storing it in tanks and cisterns. Equally, ordinary water users continually practise their own forms of demand management: sacrificing cooked meals so as to minimise the washing-up, letting the laundry pile up until the water returns, taking showers elsewhere at the homes of those with water. Where in Israel it is state institutions that oversee fluctuations in supply and consumption levels (recall the enormous variations in water use that they administered during the early 1990s), in the southern West Bank it is families, individuals and especially women who do the ultimate work of governing domestic water supply and demand. Most southern West Bank Palestinians are in short not passive consumers of water – conforming to a neo-liberal institutional ideal – but are active, inventive and even expert managers of their own water supplies.[18]

Fourthly, the coping practices detailed here are intimately bound up with everyday social interactions, and as such attain degrees of meaning and symbolic significance. However, just as material practices display a great deal of variety, so too do the meaning and significance of water and water shortages. In Dheisheh, for instance, shortages are only coped with through the constant giving, taking and sharing of water, such that shortages become a source of and reason for sociality. The return of piped water to the foot of the camp instigates an almost carnivalesque occasion, during which washing and carrying are accompanied by shouts of joy and a constant buzz of relieved chatter. Similarly, Ayn Artas and other West Bank springs are, for many, sites for play and interaction. Water shortages can also have a degree of political meaning. In Dheisheh, shortages constitute everyday material evidence of occupation and dispossession. Filling up his water containers at a nearby spring, one young Palestinian man told me, 'this is good for us, good to learn; it teaches people how to get things from the difficult life... you will do anything to make you human; if you have a satellite, everything you need, you will stop thinking about 1948'. People observe time and again that they do and will share their water, 'until the last drop'.[19] Others, unsurprisingly, see things totally differently: for them, water shortages bring shame and ritual humiliation, making 'you feel like you're not human,' as one Palestinian woman told me.[20] Meanings are of course important, but they are variable and flexible, and they arise and are reproduced only through the contexts of social interaction. They are not culturally determined by homogenously Middle Eastern symbolic structures and belief systems.

In closing on this note my intention is not to romanticise the resistant activity of ordinary Palestinians, nor indeed to suggest that Palestinian water shortages are not as severe as they might at first seem. I have little doubt that the vast majority of Palestinians living in Bethlehem would be more than glad to see an improvement in their water supplies; and neither do I doubt that the water crisis in the Occupied Territories has severe economic, political and health effects.[21] At a household level, the burden of adapting to water shortages generally falls to a much greater extent on women than it does on men, inevitably affirming the already strongly patriarchal nature of Palestinian society. More widely, the practical demands of 'getting by' in the face of shortages results in sometimes violent conflicts between households, streets, communities, municipalities, and between them and central institutions. There is no romance in doing without piped water supplies for several months on end. What there is, however, is a degree of agency – the fact that despite all the powerful structures and the weight of history that lie behind present-day water shortages in Hebron and Quasiba, people's lives are not just structurally determined. People suffer, no doubt, but they also adapt, cope and generally get by.

Conclusion

My aim in this book has been to develop an original and critical analysis of the nature and causes of the ongoing Palestinian water crisis. I have sought in particular to argue that this crisis needs to be approached with an eye to a range of different levels and scales – from the long-historical patterns of state formation and development within which water crises emerge, to the micro-scale practices through which people adapt to water shortages in the course of everyday interaction. I have sought, also, to suggest that it makes little sense to discuss technical and political issues in isolation from one another since the two are always inextricably linked – whether in the form of established and changing patterns of political economy, in the form of those technological networks through which discriminatory policies are effected, in the form of political disputes over scientific truths, of pitched battles taking place over control of water valves, or of the household level gender politics involved in coping with water shortages.

More specifically I have argued that the roots of the Israeli–Palestinian water conflict lie in the specific form of the colonial encounter between the Zionist movement and the Palestinians, and the characteristically Zionist and Israeli institutions, and patterns of state-society relations, to which this encounter gave rise. The results of these were that water came to be of paramount importance to the Israeli state and society, and simultaneously that Israel developed strong institutional forms that were well able to establish a high level of internal control over water resources, and able also to direct these resources towards Jewish Israeli nation-building purposes. Following 1967, Israel's colonial drive to establish settlements throughout the land of Palestine was extended, and its existing discriminatory and indeed apartheid treatment of its Palestinian minority was deepened and radicalised. The result, as far as water is concerned, was the construction of a water supply network that ensured that water was diverted to Israeli settlements at the expense of Palestinian communities, the creation of a dual legal regime that limited Palestinian water use in order to maximise the amount of water available for Jewish Israeli society, and the formation of a client system that enabled the Israeli state to control the West Bank's water resources while having only minimal administrative contact with the occupied Palestinian population.

Under Oslo not that much changed. Israeli colonisation continued, but a formal quisling authority was now established, its primary function being to do Israel's security bidding in the West Bank and Gaza. Israel retained control over all vital water resources. A new and this time formally legitimised client system was set up for the management of the local Palestinian water sector and the international community was invited to fund its economic reconstruction, albeit within terms defined by the Israeli authorities. The Palestinians were granted the right to develop the West Bank's Eastern Aquifer on the seemingly erroneous assumption that plenty of additional waters were available from there, possibly to disastrous effect. The PLO was willing to put its name to all of this out of political weakness and desperation, because decision-making was concentrated in the hands of Yasser Arafat, because it made little use of technical expertise, and in the vague hope of Palestinian statehood.

The PA system established on the back of Oslo has been fragmented, security-led, donor-dependent and poorly institutionalised. The result has been that in stark contrast with Israel's centralised state water institutions, the Palestinian Water Authority and West Bank Water Department have faced recurrent difficulties in controlling and administering their water sector. In consequence, Palestinian society has generally not seen the benefits expected even from the limited terms of the Oslo Accords. Much of the work of supply and demand management that in Israel is conducted by state institutions, in the West Bank has to be performed by individuals, households and local communities. Since September 2000, and with the increasing fracturing of the Palestinian Authority, this has no doubt become even more the case. Since then, in the midst of much-heightened levels of Israeli repression, almost continual closure, and a growing economic and humanitarian disaster, many infrastructures have deteriorated and been damaged, the work of maintaining and administering the water sector has become near impossible, and more and more Palestinians have been forced to endure water shortages. Almost ten years on, the hoped-for benefits of the Oslo process have come to very little.

With this all in mind, we need finally to consider three issues: the reasons for the collapse of the Oslo process; the likelihood that the Palestinians' water problems will be resolved in the near future; and the policy-making implications of the foregoing analysis. It is to these concluding questions that we now turn.

THE FAILED 'REMOTE CONTROL OCCUPATION'[1]

At the outset of the Oslo process, both Rabin and Peres were opposed to the idea of Palestinian statehood, as we have already seen. Nonetheless, it did not take long for this to change, with the Labor Party dropping its official opposition to Palestinian statehood prior to the 1996 elections, and senior Likud politicians including Ariel Sharon also coming to accept the idea. This did not necessarily imply a great deal, however. Writing in 1996, before the Israeli Labor Party had changed its stance on Palestinian statehood, Finkelstein argued that if 'the South African precedent is any guide, Israel will eventually grant the Palestinians full independence within the patchwork of areas of "self-rule" adumbrated in Oslo II'.[2] Finkelstein observes that the South African Bantustan of Transkei was granted formal independence by South Africa in 1976, and that in its formal relations with South Africa, if not in the fact that it received no international recognition, 'it did enjoy the same legal status as any other state'.[3] This assessment will probably in the end prove to have been over-pessimistic. The area under full or partial PA control was, after all, extended following the Wye and Sharm-el-Sheikh agreements, and it still seems highly likely that a Palestinian state will at some point be established on a goodly part of the West Bank and Gaza. Nonetheless, Finkelstein's warning against the dangers of taking juridical and discursive changes at face value is well taken. Irrespective of whether Israeli–Palestinian water relations are designated 'co-operative' or not, the key issue is how different powers and responsibilities are actually distributed. Equally as Chomsky notes, 'whether the US and Israel decide to call the cantons they allow the PLO to "govern" a "state" or something else – perhaps "fried chicken" as [Binyamin Netanyahu's spokesman] David Bar-Illan elegantly suggested – the results are likely to resemble the Bantustan model'.[4] The important question is not whether the Palestinians will attain nominal statehood, but what the powers and extent of this future Palestinian state will be.

With these initial thoughts in mind, we can now consider the reasons for the failure of the Oslo final status negotiations and the subsequent violence in Israel and the Territories. Final status discussions on all of the issues deferred by the Oslo agreements – the territorial and juridical powers of the future Palestinian entity, as well as the fate of Israeli settlements, Jerusalem and Palestinian refugees – took place during 2000 and early 2001, first in Israel and the West Bank, then in Camp David, Maryland, and thereafter back in the Middle East, these including negotiations on water issues.[5] Their failure has often been blamed on Palestinian intransigence, specifically Yasser Arafat's failure to accept Ehud Barak's unprecedented generosity at Clinton's Camp David summit. The Clinton administration, for one, placed the responsibility for this failure squarely on Arafat's shoulders, arguing that, when it came to the crunch, the PA Chairman was unwilling to 'go the extra mile for peace'.[6] Since then it has become standard pro-Israeli fare that Arafat spurned a golden Israeli offer, opting instead to resort once again to terrorism – this being in keeping with Abba Eban's famous characterisation of the Palestinians as having 'never missed an opportunity to miss an opportunity'. The truth of the matter is, however, quite different. While according to most reports Barak offered to cede 90 per cent or more of the West Bank to the Palestinians, even members of his cabinet doubted whether this offer did indeed exist.[7] Palestinian negotiators at Camp David were presented only with vague ideas, not detailed proposals, insufficient preparatory work having preceded the negotiations; there was no direct negotiation between Barak and Arafat; most ideas were passed, confusingly, through American mediators, such that it was often unclear to the Palestinians whether they were being handed US or Israeli proposals; and 'strictly speaking,' as Malley and Agha conclude, 'there never was an Israeli offer'.[8] Finally, even if the 90 per cent figure was indeed an accurate reflection of Barak's position, then under these terms all major settlement blocs in the West Bank would have been annexed into Israel, including in the southern West Bank, Ma'aleh Adumin to the east of Jerusalem, the Gush Etzion settlement bloc between Bethlehem and Hebron, and even Kiryat Arba.[9] The 'independent state of Palestine' would have comprised a series of Bantustan-like non-contiguous enclaves – which would hardly have represented either a just resolution to the conflict, or a workable basis for Palestinian statehood.

A second problem with the standard pro-Israeli myth is that negotiations continued after Camp David, despite the Palestinians' supposed resort to terrorism. If insufficient preparation had gone into the earlier Camp David talks, then these later negotiations were much more serious and more productive. At the conclusion to these talks, in January 2001, the two sides publicly declared that they had 'never been closer to reaching an agreement' – an assessment that was confirmed by EU Envoy Miguel Moratinos's informed account of the final Taba negotiations.[10] At these, both sides presented maps of what the Final Status West Bank might look like, both accepted the principle of a 'land swap' (such that Israel would annex major settlement blocs but compensate the Palestinians with land from within 1948 Israel), and both exchanged ideas on all outstanding issues. The talks came to a halt not because of Palestinian intransigence, but because of the election of Ariel Sharon. Nonetheless, it is from these points that any future two-state resolution to the Israeli–Palestinian conflict will have to proceed.

In the meantime, of course, violence had broken out across Israel and the Territories. Prompted by Ariel Sharon's incendiary visit to the Harem Al-Sharif, or Temple Mount, in Jerusalem of 28 September 2000, protests and counter-violence quickly spread – leaving seven Palestinians dead in Jerusalem the following day and culminating, a year and a half later, in assassinations, suicide bombings and Israel's re-occupation of Palestinian cities across the West Bank. The Palestinians, it is often said, opted to ditch negotiations in favour of terror. Yet as the authoritative report of the Mitchell Fact-Finding Committee made clear, the Al-Aqsa intifada began as a series of confrontations between largely unarmed Palestinians and armed Israeli security forces resorting to excessive and lethal force; Mitchell, moreover, found 'no basis on which to conclude that there was a deliberate campaign by the PA to initiate a campaign of violence'.[11] To the contrary, the intifada was an 'organised response by the Young Guard in the Palestinian national movement not only to Sharon's visit al Haram al Sharif and to the failure of the peace process to end Israeli occupation, but also to the failure of the PLO's Old Guard to lead the Palestinian process of independence, state building, and governance'.[12] Put another way, the intifada represented a response by those activist insiders who had been politically marginalised under Oslo, who were disillusioned with the terms and (lack of) direction of the Israel–Arafat agreement, and who sought, as during the first intifada, to take a lead in resisting Israel's ongoing occupation. Despite constant Israeli assertions to the contrary, there is no evidence directly implicating Yasser Arafat in terrorist attacks within Israel.[13] While it can quite cogently be maintained that Arafat has not done enough to curtail Palestinian violence – since to do so would involve loosening his own grip on power, and threatening still further his own declining domestic legitimacy – the argument that the Oslo peace process broke down because of Palestinian intransigence, or due to a proclivity for terror, is quite simply mistaken.

Irrespective of where the violence was coming from, however, the PA under Yasser Arafat was no longer fulfilling the security functions for which it had been established by Rabin, and it was for this reason that he was declared 'irrelevant', first by Ariel Sharon and later by the Bush Administration. Unable to rely upon its client police force in Gaza and the West Bank, Israel has for the most part, since March 2002, destroyed the Palestinians' policing capability (most clearly symbolised in the destruction of Jibril Rajoub's Preventative Security compound in Ramallah in April 2002), and re-established direct occupation. The idea of the PA has been preserved, though, primarily because it continues to serve 'as a rhetorical buffer obfuscating the true relations of forces on the ground', and because it grants Israel's occupation a degree of legitimacy and hence defers the costs of administering the Territories to the international community.[14] Over issues such as water, as we have seen, Israeli–Palestinian co-operation continues as before.

A generous assessment of the Oslo process might be that it has collapsed because the PLO's concessions were all to be made at the beginning, but most of Israel's at the end. Malley and Agha contend, for instance, that Ehud Barak's strategy was one of declining to make concessions to the Palestinians during interim negotiations in order to store up political capital for final status negotiations.[15] Whether this was the case or not, by September 2000 Israel had provided little material evidence of any such intent. In the meantime, the obvious fact that the occupation was ongoing, the continued growth and extension of settlements, the continued expropriation of Palestinian land, the mounting evidence of PA corruption and ineffectiveness, and the fact that the Palestinians were now being blamed for not making concessions – all of this fed growing popular resentment against Israel and increasing disenchantment with the PA leadership. It is within this context that events since September 2000 need to be understood: as the culmination of a process which was structured in accordance with Israeli power and Israeli interests, and which showed little sign of bringing Palestinian dispossession and statelessness to a just or meaningful end. It is within this context, moreover, that the Palestinian water crisis also needs to be explained.

RESOLVING THE WATER CRISIS

The immediate causes of the current water crisis in the Palestinian territories are Israeli closure and military occupation – which, as we saw in the introduction, are creating an economic and humanitarian crisis, are resulting in the destruction and decay of water supply infrastructures, and are making it very difficult to transport water to isolated villages. In the short term, this crisis is only being held at bay by the work of emergency relief organisations. For a more meaningful resolution, what is required is the end of the collective economic and political punishment of the Palestinian population.

In the longer term, though, what is the chance that the West Bank water crisis will soon be overcome? My personal assessment is pessimistic since, for this crisis to be resolved, the Palestinians will require both a larger and more equitable share of the West Bank's water resources, and a strong administrative state with the capacity to govern resources, supplies and demand. In neither of these respects are the signs particularly encouraging.

Israel is currently adamant that it will 'not allow them [the PA] to drill in the Western Aquifer, or the North-eastern Aquifer'.[16] It is possible that Israel will make some minor concessions on these aquifers during final status negotiations, though rather more likely that Arafat, if he ever gets to negotiate final status terms, will be the one making concessions on water in a bid to wrest territorial concessions from Israel – since extending control of territory, rather than promoting development of the Palestinian economy or establishing conditions for institution-building has been Arafat's focus throughout the Oslo period. As for the Eastern Aquifer, it is possible that the PWA will at some point be granted control over the Mekorot wells that tap into it, and the supply lines leading from them. Israel, as noted earlier, is currently constructing new lines to Israeli settlements, justifying this on the grounds that once these settlements are fully supplied from within Israel, the Eastern Aquifer wells will be 'related solely to Palestinians', and will thus, under the terms of the Oslo II water accords, come under PA control. While unlikely to occur before any final status agreement (for obvious tactical reasons), it does seem likely that at some point the PA will be granted full or near-full rights over the Eastern Aquifer. However, to this conclusion an important qualification should be added, one that returns us to chapter 5 – that the safe yield of the Eastern Aquifer appears to be far below the 172 mcmy suggested in the Oslo II Agreement. If this were the only additional source of water supplies granted to the West Bank's Palestinians as part of a final status deal, then it would seem likely that they would continue to face water shortages, and may even end up courting environmental disaster.

With regard to the possibility of the Palestinians developing strong state water institutions, the prospects are equally gloomy. The Israeli case is a useful exemplar here. Israel's infrastructural and institutional capabilities are the products, as we have seen, of a specific sort of colonial encounter, this having given rise to centralised pre-state and state institutions devoted to nation-building and the extension of territorial control. Its capabilities are also the product of economic power. Without these, Israel would not have been able to construct such an integrated national water network, and would not be able to contemplate large-scale desalination. Any future Palestinian state is likely to be very different in its capabilities. In straightforward economic terms, while the West Bank had, prior to September 2000, a total GDP of around $3.3 billion, and a per capita GDP of $2050, Israeli GDP was $105.4 billion, that is $18,300 per capita.[17] Institutionally, moreover, the patterns established under Oslo will be likely to have lasting effects. As Hillel Frisch writes in his excellent comparative analysis of Israeli and Palestinian state-building, the

> Israeli experience is an excellent example of how pre-independence state-building patterns can predict the character of the state after independence. Only in 1977, nearly thirty years after the establishment of the state, did the right-wing Likud party take the reins of power from the hegemonic state-building political elite. In the emerging Palestinian entity, the legacy of Arafat's neopatrimonialism will probably be even more considerable. As in the Israeli case, Palestinian patterns of state-building are likely, after the assumption of self-government, to predict the type of state consolidation for a considerable period of time.[18]

If this is indeed the case, it is hard to imagine that a future Palestinian state would able to finance large-scale water and wastewater facilities, to maintain modern water tariff systems, or to enforce payment for expensive Eastern Aquifer or desalinated water. The Palestinian system is likely to remain both economically and institutionally weak, without the administrative capabilities to produce and distribute water, to regulate demand, or to resolve the domestic dimensions of water crisis.

Whether even these developments will eventually come to pass, depends in large measure on developments within Israeli society. It was noted previously that Israeli society has fragmented since the 1970s, with the Labor Zionist consensus that dominated the early years of statehood having increasingly come under challenge both from religious-nationalist and liberalising movements. Within the water sector, while policy-making used to be dominated by agricultural and nation-building agendas, since the early 1990s it has increasingly borne the marks of economic liberalisation. Mekorot and Tahal have both been restructured, the former becoming an autonomous cost-centre which now operates without government subsidies, and the latter having been wholly privatised.[19] For the first time, in 1990 a Water Commissioner was appointed who was not from the agricultural establishment.[20] In 1996, the Water Commission itself was transferred from the Ministry of Agriculture into a new Ministry of Infrastructure. Then in 1997, a national commission made a series of significant recommendations, urging amongst other things the elimination of subsidies, the raising of tariffs and the privatisation and break-up of Mekorot.[21] These developments have been halting and not without their contradictions. Nonetheless, their impact on Israeli–Palestinian water relations might, as Eran Feitelson argues, be great, since the possibility of the Palestinians being granted a greater share of regional water resources depends to a large extent on how Israel interprets its water needs.[22] If the agricultural sector declines in influence, and if water is allocated according to more purely economic criteria, then more water is likely to become available for concession to the Palestinians. The route to the Palestinians being granted access to the Western and North-eastern Aquifers, or even to the Jordan River, in sum, lies above all in the liberalisation of the Israeli economy, polity and society. The tragedy is that, quite apart from their cost in human lives, terrorist attacks on Israeli cafes and supermarkets end up affirming religious, nationalist and other right-wing perspectives at the expense of liberal ones – and in the process push both the possibility of statehood, and the possibility of being granted increased water supplies, still further into the distance.

POLICY IMPLICATIONS

At the outset of this study I was explicitly critical of problem-solving approaches to the analysis of Middle Eastern water issues, arguing that while they tend to be strong on description and prescription, they are typically quite weak in explanatory terms. Being problem-solving in orientation, and being so enmeshed in the power/knowledge apparatuses of state and inter-state institutions, they tend to repeatedly invoke the distinction between technical and political factors, to ignore structural questions, and to generally understate issues that state institutions would also rather not talk about. Key amongst my aims in this study has been to draw attention to some of these often marginalised issues and, more broadly, to develop an explanatory account that tries to problematise problem-solving, and refuses to make any straightforwardly technical prescriptive gestures. To this extent, the overall tenor of this book has been to argue against concluding with a neat list of policy implications. That said, it behoves me in conclusion to make a few brief policy-relevant comments, however uncertain and technically unsatisfying these may be.

It was argued in chapter 1 that water crises need to be understood in general as problems of political economy, whose roots are never either Malthusian, or simply technical or political. I have also argued that water problems need to be analysed as arising within a world systemic framework, and as the products of specific patterns of state-society relations. The implication of these structuralist political economic arguments is that the Palestinians' water problems will never be resolved simply through them being granted a more equitable share of regional water resources. What they require more broadly is the political economic and institutional capacity to manage resources across space and time, and to make use of them as effectively as possible. This means that they will need water institutions which can effectively intervene in, regulate and mould Palestinian society, and which are not beholden to the interests and agendas of local 'strongmen', as well as an economy that is strong enough to enable sufficient and indeed continual investment in the water sector. Unless the institutions of the Palestinian proto-state and state are to continue being dependent on international donors, they will at some point have to develop the economic capacity to recycle and desalinate water. And to achieve this what they will need, above all, is some platform for sustainable economic and especially manufacturing growth. In recent years these have been retarded not only by the military face of occupation, but also by the pattern of dependent development to which the Occupied Territories have been made subject. To open up the possibility of sustained economic development, in short, the Palestinians need to be freed from this relationship of dependency with Israel. Given this, I would venture to contend that it will be the economic terms of any future final status agreement with Israel, rather than those terms that specifically relate to water, which will thereafter have the greater bearing on the Palestinians' hydro-political future.

Here and in much of the foregoing analysis I have emphasised the importance of institutional structures and capabilities in managing scarce water resources and supplies. The potential implications of this for water policy (and for development work within the water sector) are several, and too intricate to discuss at any great length. Instead I simply pose a number of questions regarding the appropriate sites for donor funding; the neo-liberal preference for private sector contractors and water utilities; the value of donors' seeming obsession with plans, papers and policy documents; and regarding the contribution of international donors when operating within unjust conflict situations. We consider these briefly in turn; I pose them only as questions, and without providing, or being able to provide, any firm answers.

- Some donors take the view that 'the goal of working to make municipalities more accountable and to promote consensual municipal governance is best served by having primary contractual relationships with individual municipalities'.[23] Others have sought to route all their development assistance through central state institutions. Given the existence of conflicts between central and local authorities, which of these represents the more appropriate site for donor funding? While strengthening central institutions might be a developmental priority, these institutions often focus their attentions on major population centres (especially Bethlehem in the case of the West Bank) and on large-scale projects, without much regard for smaller towns and villages. Local communities, for their part, may simply use donor funds as a political tool in struggles with national and other local protagonists.
- Does private sector contracting and sub-contracting represent an appropriate model for development work in areas such as the southern West Bank? Clearly this model is motivated by certain core capitalist interests and leads to the return of a high proportion of donor aid to the developed North. Quite apart from that, however, does it not but complicate the work of water institutions in attempting to govern the water sector?
- How can central regulatory agencies be most supported in developing institutional structures and viable plans and policies? Do international consultants provide much-needed expertise? Or do their typical lack of country-specific knowledge, the danger of duplication and consistency, and the inevitable problems of ownership and consensus-formation mean that institutional development is best left just to those on the inside? More generally, are institutions that have been constructed almost solely on the basis of international aid money likely to be sustainable in the medium- and long-term?

- Finally, the post-Oslo development effort has arguably been premised on a transfer of burdens and responsibilities from Israel to the international donor community. Given that this donor effort was initiated at a time when Israeli leaders remained set against Palestinian statehood, did this not simply legitimise the occupation, and allow Israel freedom from a set of burdens that, through its de-development of the Palestinian territories, it had itself created? Have not international donors since continued in this role, even while Israel has been going about its re-occupation of the West Bank? Are there not difficult questions to be asked here about whether international donors are in practice propping up the Israeli occupation?

Development work, like all other forms of water-related problem-solving, can either contribute to ameliorating water problems, make them worse, or more typically end up having a range of variable social and political effects. This study has sought to highlight some of the complexities involved in doing technical work at the inevitable intersection between problem-solving, politics and society. But it has also sought to emphasise that, when all is said and done, water crises have structural roots in patterns of political economy and state-society relations that are well worth taking into account. As the Oslo process demonstrates as a whole, problem-solving without structural change can be a recipe for disaster.

Appendix: Interviews and Meetings

The following is a list of all the people who have helped me with my research in Israel, the West Bank and Gaza (not including the many 'ordinary' people who also helped me greatly). Many of these did not involve formal interviews. The positions given for each of the interviewees refer to those held when I first met them. Thanks to all of them. None bear any blame for anything in this text, except perhaps where they are specifically cited.

Abbas, Ziad; Dheisheh Community Centre, Bethlehem (30 March 1998).
Abdul Latif, Mohammed; PARC, Ramallah (4 July 1998).
Abu-Deyah, Mahmud; Senior Engineer, WSSA (16 July and 25 August 1998, 19 August 1999, 4 June 2002).
Albert, Jeff; School of Forestry and Environmental Studies, Yale University (various 1998 and 1999).
Aliewi, Amjad; PHG (16 April 1998, 1 June 2002).
Alkan, David; A.D. Systems, and former Head of Planning, Southern and Central Areas, Water Commission, Israel (various 1998 and 1999).
Areikat, Maen Rashid; Director General, Negotiations Affairs Department, PA (26 February 1999).
Arlosoroff, Saul; Zafrir Weinstein Engineers and Consultants Ltd, Tel Aviv (5 April 1998).
Asa'd, Abdelkarim; General Manager, Jerusalem Water Undertaking, Ramallah District (20 July 1998).
Assaf, Karen; PWA (13 July 1998, 17 August 1999, 1 June 2002).
Atfield, Katie; ICRC, Jerusalem (5 June 2002).
Attal, Nabil; Engineer, Dura Municipality, Dura (10 July 1998, 23 August 1999).
Attili, Shaddad; Negotiation Affairs Department, PA (2 June 2002).
Aviram, Ram; Director, Multilateral Peace Talks Coordination and Water Issues, Israeli Ministry of Foreign Affairs (19 July 1998).
Awad, Qasem; Resident Engineer, CDM/Morganti, Bethlehem (26 August 1998 and 24 August 1999).
Bargouthi, Ihab; PWA (12 August 1998, 18 August 1999, 2 June 2002).
Bashir, Basema; Manager, Hydrological Information System, PWA (4 April 1998).

Besser, Moshe; Manager of Logistics and Quality Controller of Avocado, Fruit Board of Israel (8 April 1998).
Bromberg, Gidon; Secretary General, Ecopeace, East Jerusalem (9 April 1998).
Carmi, Natasha; MOPIC, PA (31 March 1998).
Coulter, David; Save the Children Fund, London (8 June 1998).
Daibes, Fadia; Deputy Programme Director, Norwegian Institutional Co-operation Programme, PWA (various 1998 and 1999).
Dajani, Ibrahim; MOPIC, PA (various 1998 and 1999).
Eikenaar, Esme; Ecopeace, East Jerusalem (various 1998).
Feitelson, Eran; Department of Geography, Hebrew University of Jerusalem (16 April 1999).
Foster, Annie; Programme Director, Save the Children Federation, A-Ram (17 June 1998).
Gillad, Dvor; IHS, Jerusalem (16 August and 23 August 1998).
Golani, Ze'ev; Mekorot (6 August 1998).
Guttman, Yossi; Senior Hydrologist and Hydro-Geologist, Tahal (4 August 1998).
Haddad, Marwan; Dean, Faculty of Engineering, An-Najah University, Nablus (8 August 1998).
Halasa, Malu; freelance journalist, London (various 1998 and 1999).
Hamis, Lesley; former UNDP employee, Jerusalem (various 1998 and 1999).
Hamzeh-Muhaisen, Muna; freelance journalist, Dheisheh Refugee Camp, Bethlehem (17 August 1998).
Haugan, Frank; Mapping/GIS Advisor, Physical Planning and Institution Building Project, MOPIC, PA (various 1998).
Hauge, Olaf; Project Manager, Physical Planning and Institution Building Project, MOPIC, PA (various 1998).
Hosh, Leonardo; Programme Officer, Global Environment Facility, UNDP, East Jerusalem (24 June 1998).
Isaac, Jad; Director, ARIJ, Bethlehem (various 1998, 1999, 2002).
Ismail, Mahmoud; Hydro-Geologist, PWA, Orabi St, Gaza (15 April 1998).
Jaas, Mohammed; WBWD (26 August 1998, 18 August 1999, 1 June 2002).
Jabarin, Idris Shaker; Head, Village Council, Duwarra (various 1998).
Jaber, Ibrahim; Economist, PWA, Orabi St, Gaza (15 April 1998).
Jarrar, Ayman; PWA (17 August 1999, 1 June 2002).
Al-Jayyousi, Anan; Director, WESC, An-Najah University, Nablus (14 April and 22 August 1998).
Kantor, Schmuel; Co-ordinator Special Duties, Water Commissioner, and Senior Advisor, Mekoroth, Tel Aviv (13 August 1998).
Al-Khateeb, Nader; WEDO (6 April and 12 July 1998).
Khatib, Ghassan; JMCC, East Jerusalem (11 April 1998, 15 August 1999).
Kinnarty, Noah; Ministry of Defence, Israel, and former Israeli Co-ordinator of the JWC (28 July 1998).
Kittani, Hazem; PWA, El-Bireh (22 July 1998).

Al-Labadi, Ali; Managing Director, Universal Engineering Consulting, Amman (31 March 1998).

Laham, Mohammed; Director, Camp Services Committee, Dheisheh Refugee Camp, Bethlehem (various 1998 and 1999).

Laham, Tigritte; Palestinian Ministry of the Environment, Ramallah (various 1998 and 1999).

Levitte, Dov; Geological Survey of Israel, Jerusalem (18 June 1998).

Maas, Lucianne; Save the Children Federation, Halhoul (various 1998).

Meyer, Jane; Programme Management Officer, UNDP, East Jerusalem (24 July 1998, 17 August 1999).

Morris, James; James W. Morris and Associates, Inc., Standish, Maine (12 July 1998).

Nafe, Ahmad; GIS Specialist, PWA, El-Bireh (various 1998 and 1999).

El-Najjar, Safeya Hosney; Director, PWA, Orabi St, Gaza (15 April 1998).

Nassereddin, Taher; Head, WBWD, Bet El (12 April and 15 August 1998).

Naudy, Andre; Vivendi and General Des Eaux, Halhoul (25 August 1999).

Nuseibi, Mustapha; Head, Hydrological Monitoring Department, WBWD, Bet El (27 June 1998).

Qannam, Ziad; WSERU, Bethlehem University, Bethlehem (30 March 1998).

Qdaih, Osama; Khuza'a Permaculture Centre, Khuza'a, Khan Younis, Gaza (15 April 1998).

Rabbo, Alfred Abad; WSERU, Bethlehem University, Bethlehem (30 March 1998).

Refa'ai, Samira; Save the Children Federation, Halhoul (various 1998 and 1999).

Rihan, Salam; Civil Engineer, PHG (5 July 1998).

Sansur, Ramzi; Director, Centre for Environmental and Occupational Health Sciences, Birzeit University, Birzeit (2 April 1998).

Sbeih, Mohammad; Irrigation Projects Co-ordinator, ANERA, East Jerusalem (16 July 1998).

Scarpa, David; WSERU, Bethlehem University, Bethlehem (various 1998, 1999, 2002).

Schwarz, Joshua; Senior Hydro-Geologist, Tahal (24 August 1999).

Shaheen, Husayn; Camp Services Director, Dheisheh Refugee Camp, Bethlehem (16 August 1999).

Shalan, Walid; Save the Children Federation, Halhoul (various 1998 and 1999).

Sharif, Nabil; Chairman, PWA (11 February 1998).

Shuval, Hillel; School of Applied Science and Technology, Hebrew University of Jerusalem (28 June 1998).

El-Surdi, Hassan; Environmental Health Educator, PWA, Orabi St, Gaza (15 April 1998).

Tahboub, Hiba; World Bank, Jerusalem (9 July 1998).

Tamimi, Abdul Rahman; General Director, PHG, A-Ram (19 April 1998).

Usher, Graham; freelance journalist, Jerusalem (3 April 1998).

Weighill, Marie-Louise, Save the Children UK, Jerusalem (4 June 2002).

Yaqoub, Iyad; PWA, El-Bireh (various 1998 and 1999).
Yizraeli, Moshe; Consultant to the Water Commission (21 July 1998).
Zahrer, Bader; MOPIC, PA, A-Ram (31 March 1998).
Zarour, Hisham; Head, Water Resources and Planning Department, PWA (various 1998, 1999, 2002).
Zaslavsky, Shmuel; Chief Engineer, Mekoroth, Ramle (24 August 1998).
Zayad, Omar; PWA, El-Bireh (17 August 1999).
Zir, Imad; Hebron Municipality (5 August 1998).

Notes on the Text

Notes on Introduction

1 Amnesty International, 'Israel and the occupied territories: the heavy price of Israeli incursions' (14 April 2002).
2 Palestinian NGO Emergency Initiative in Jerusalem, 'Report on the destruction to Palestinian governmental institutions in Ramallah caused by IDF forces between March 29 and April 21, 2002' (22 April 2002); Amos Harel, 'IDF admits "ugly vandalism" against Palestinian property', *Ha'aretz* (30 April 2002).
3 Palestinians NGO Emergency Initiative in Jerusalem, 'Destruction of non-governmental organizations in Ramallah caused by IDF forces between March 29 and April 21, 2002' (22 April 2002).
4 Local Aid Co-ordination Committee Co-Chairs, 'Damage to civilian infrastructure and institutions in the West Bank estimated at US$361 million', press release (15 May 2002). The latter figure, for 2001, is from World Bank, 'Fifteen months: intifada, closures and Palestinian economic crisis – an assessment' (March 2002), p. 25.
5 E-WaSH, 'Nablus water situation', internal report (14 April 2002).
6 Amira Hass, '25,000 lack water in Ramallah', *Ha'aretz* (2 April 2002).
7 Janine di Giovanni, 'Children scream for water in the "city of bombers"', *Times* (London) (9 April 2002).
8 Oxfam, 'An urgent call to address the human costs of the Israeli–Palestinian conflict', Oxfam briefing note (4 April 2002).
9 Mohammed Daraghmeh, 'A scarred Nablus', *Palestine Report*, vol. 8, no 43 (24 April 2000).
10 E-WaSH, 'Tulkarm water situation', internal report (14 April 2002).
11 E-WaSH, 'Ramallah water situation', internal report (14 April 2002).
12 Interview with Ihab Bargouthi, Palestinian Water Authority (2 June 2002).
13 Local Aid Co-ordination Committee Co-Chairs, 'Damage to civilian infrastructure' (April 2002).
14 Phil Reeves, 'New barriers widen gulf on West Bank', *Independent* (26 May 2002).
15 On the worsening humanitarian situation see for instance USAID, 'Preliminary findings of the nutritional assessment and sentinel surveillance system for the West Bank and Gaza' (5 August 2002).

16 World Bank, 'Fifteen months: intifada, closures and Palestinian economic crisis – an assessment' (March 2002), p. vi.
17 Ibid.
18 Ibid.
19 Ibid., p. v.
20 Interview with Mahmud Abu-Deyah, Senior Engineer, WSSA, Bethlehem (4 June 2002).
21 Interview with Katie Atfield, ICRC, Jerusalem (5 June 2002).
22 Amira Hass, 'Cut and dried', *Ha'aretz* (31 July 1998); Hass, 'Sharon says PA excuses are all wet', *Ha'aretz* (19 August 1998); D. Jehl, 'Water divided haves from have-nots in West Bank', *New York Times* (15 August 1998).
23 Interviews with Mahmud Abu-Deyah (16 July and 25 August 1998, 19 August 1999).
24 GTZ, 'Middle East regional study on water supply and demand development, concluding report' (February 1998), p. 7. This figure for 'gross' supplies does not include water lost through leakage and theft; 'net' supplies, those actually received by Palestinians, would perhaps have been only two-thirds of this figure.
25 GTZ, 'Middle East regional study on water supply and demand development, phase 1 report' (August 1996), p. S-4.
26 This figure is often attributed to the World Health Organisation, but the 'WHO has never established any minimum domestic water requirement' (Anne-Marie Pfister, WHO, personal communication, 29 August 2000). Nonetheless the figure of 100 cmy does emphasise the extent of the water shortages in the West Bank and Gaza.
27 Israel and the PLO, 'Declaration of principles on interim self-government arrangements' (Washington DC, 13 September 1993).
28 White House press release, 'Remarks by President Clinton *et al*' (13 September 1993).
29 Rex Brynen, 'International aid to the West Bank and Gaza: a primer', *Journal of Palestine Studies*, vol. 25, no 2 (1996), p. 46; Brynen, *A Very Political Economy: Peacebuilding and Foreign Aid in the West Bank and Gaza* (Washington DC: US Institute for Peace, 2000), p. 78.
30 World Bank, *Developing the Occupied Territories: An Investment in Peace*, vol. 1: *Overview* (Washington DC: World Bank, 1993), p. v; World Bank, *Emergency Assistance Programme for the Occupied Territories* (Washington DC: World Bank, 1994), pp. 8–9; interview with Hiba Tahboub, World Bank, Jerusalem (9 July 1998).
31 USAID, 'US assistance to the West Bank and Gaza', at http://www.info.usaid.gov/regions/ane/newpages/one_pagers/wbg.htm. USAID, 'USAID: funding for the West Bank and Gaza', at http://www.info.usaid.gov/wbg/funding.htm.
32 Alwyn Rouyer, *Turning Water into Politics: The Water Issue in the Palestinian–Israeli Conflict* (London: Macmillan, 2000), p. 229.

33 Chomsky, *Fateful Triangle: The United States, Israel and the Palestinians*, 2nd edn (London: Pluto, 1999), ch. 10; 'Interview with Chomsky', *Z Magazine* (2 April 2002); Said, *Peace and its Discontents: Gaza–Jericho 1993–1995* (London: Vintage, 1995); Said, *The Politics of Dispossession: The Struggle for Palestinian Self-Determination 1969–1994* (London: Vintage, 1995); Said, *The End of the Peace Process: Oslo and After* (London: Granta, 2000); Said, *Power, Politics and Culture: Interviews with Edward Said*, ed. Gauri Viswanathan (New York: Pantheon, 2001); Benvenisti, *Intimate Enemies: Jews and Arabs in a Shared Land* (Berkeley: University of California Press, 1995); Bashara, *Palestine/Israel: Peace or Apartheid* (London: Zed, 2001); Usher, *Palestine in Crisis: The Struggle for Peace and Independence after Oslo*, 2nd edn (London: Pluto, 1997); Usher, *Dispatches from Palestine: The Rise and Fall of the Oslo Peace Process* (London: Pluto, 1999).

34 Horkheimer, *Critical Theory: Selected Essays* (New York: Herder and Herder, 1972); Wallerstein, *The Modern World System*, vols 1–3 (San Diego: Academic Press, 1974, 1980, 1989) and *The Capitalist World Economy: Essays* (Cambridge: Cambridge University Press, 1979); Mills, *The Sociological Imagination* (Oxford: Oxford University Press, 1959).

35 The 'levels of analysis' problem in IR is most famously articulated by J. David Singer, 'The level of analysis problem in International Relations', in Klaus Knorr and Sidney Verba (eds), *The International System: Theoretical Essays* (Princeton: Princeton University Press, 1961), pp. 77–92; and Kenneth Waltz, *Man, The State and War* (New York: Columbia University Press, 1959).

36 Wallerstein, *The Modern World System*.

37 'Theses on Feuerbach', in *Early Writings*, ed. Lucio Colletti (London: Penguin, 1975), p. 422; 'Preface to "A Contribution to the Critique of Political Economy"' in Marx and Engels, *Selected Works*, vol. 1 (Moscow: Foreign Languages Publishing, 1962), p. 363.

38 Representative examples of this Weberian tendency are Ernest Gellner, *Muslim Society* (Cambridge: Cambridge University Press, 1981) and James Bill and Robert Springborg, *Politics in the Middle East*, 4th edn (New York: Harper Collins, 1994), especially ch. 4. Weber's view of Islamic society is discussed in Bryan Turner, *Weber and Islam* (London: Routledge and Kegan Paul, 1974).

39 In this I follow Roger Owen, *State, Power and Politics in the Making of the Modern Middle East* (London: Routledge, 1992) and in particular Simon Bromley, *Rethinking Middle East Politics: State Formation and Development* (Cambridge: Polity, 1994), which develops a critique of 'culturalist' reasoning on the Middle East, as well as an excellent account and comparative analysis of state formation in the Middle East from a historical materialist perspective.

40 Robert Cox, 'Social forces, states and world orders: beyond international relations theory', *Millennium*, vol. 10, no 2 (1981), p. 128. Cox actually

writes that 'theory is always for someone...', but I would contend that his claims can be extended to cover 'knowledge' in general.

41 Ibid., pp. 128–30.

42 John Bulloch and Adel Darwish, *Water Wars: Coming Conflicts in the Middle East* (London: Victor Gollancz, 1993); Greg Shapland, *Rivers of Discord: International Water Disputes in the Middle East* (London: Hirst and Co, 1997); Tony Allan, *The Middle East Water Question: Hydropolitics and the Global Economy* (London: I.B.Tauris, 2000). Other comparative works are Mustapha Dolatyar and Tim Gray, *Water Politics in the Middle East: A Context for Conflict or Co-operation?* (London: Macmillan, 2000); Daniel Hillel, *Rivers of Eden: The Struggle for Water and the Quest for Peace in the Middle East* (New York: Oxford University Press, 1994); Nurit Kliot, *Water Resources and Conflict in the Middle East* (London: Routledge, 1994); Arnon Soffer, *Rivers of Fire: The Conflict over Water in the Middle East* (Lanham: Rowman and Littlefield, 1999); and Joyce Starr, *Covenant Over Middle East Waters: Key to World Survival* (New York: H. Holt, 1995). The above list does not include books published in languages other than English, and also omits edited collections (on which, see below).

43 Sharif Elmusa, *Water Conflict: Economics, Politics, Law and the Palestinian–Israeli Water Resources* (Washington DC: Institute for Palestine Studies, 1997); Adam Garfinkle, *War, Water and Negotiation in the Middle East: The Case of the Palestine–Syria Border, 1916–1923* (Tel Aviv: Moshe Dayan Center, Tel Aviv University, 1994); Elisha Kally and Gideon Fishelson, *Water and Peace: Water Resources and the Arab–Israeli Peace Process* (Westport: Praeger, 1993); Sarah Lees, *The Political Ecology of the Water Crisis in Israel* (Lanham: University Press of America, 1998); Steve Lonergan and David Brooks, *Watershed: The Role of Fresh Water in the Israeli–Palestinian Conflict* (Ottawa: IDRC, 1994); Miriam Lowi, *Water and Power: The Politics of a Scarce Resource in the Jordan Basin Area*, 2nd edn (Cambridge: Cambridge University Press, 1995); Rouyer, *Turning Water into Politics*; Martin Sherman, *The Politics of Water in the Middle East: An Israeli Perspective on the Hydro-Political Aspects of the Conflict* (London: Macmillan, 1998); Julie Trottier, *Hydropolitics in the West Bank and Gaza Strip* (Jerusalem: PASSIA, 1999); and Aaron Wolf, *Hydropolitics Along the Jordan River: Scarce Water and its Impact on the Arab–Israeli Conflict* (Tokyo: United Nations University Press, 1995).

44 Tony Allan (ed.), *Water, Peace and the Middle East: Negotiating Resources in the Jordan Basin* (London: I.B.Tauris, 1996); Tony Allan and Chibli Mallat (eds), *Water in the Middle East: Legal, Political and Commercial Implications* (London: I.B.Tauris, 1995); Hussein Amery and Aaron Wolf (eds), *Water in the Middle East: A Geography of Peace* (Austin: University of Texas Press, 2000); Ali Bagis (ed.), *Water as an Element of Co-operation and Development in the Middle East* (Ankara: Hacettepe University and Friedrich Nauman Foundation, 1994); Asit Biswas (ed.), *International Waters of the Middle*

East: From Euphrates-Tigris to Nile (Bombay: Oxford University Press, 1994); Jad Isaac and Hillel Shuval (eds), *Water and Peace in the Middle East* (Amsterdam: Elsevier, 1994); Peter Rogers and Peter Lydon (eds), *Water in the Arab World: Perspectives and Prognoses* (Cambridge MA: Harvard University Press, 1994); Waltina Scheumann and Manuel Schiffler (eds), *Water in the Middle East: Potential for Conflicts and Prospects for Cooperation* (Berlin: Springer, 1998); and US National Academy of Sciences, *Water for the Future: The West Bank and Gaza Strip, Israel and Jordan* (Washington DC: National Academy Press, 1999).

45 Max Weber famously defined rationalised and 'bureaucratic' organisation as the ideal typical hallmark of modern institutions; see his *Economy and Society: A Outline of an Interpretive Sociology*, vol. II (New York: Bedminster, 1968), ch. 11. Michel Foucault conceptualised the modern state as increasingly 'governmentalised' (i.e. concerned with the administrative and disciplinary governance of conduct at a distance); see his 'Governmentality', *I & C*, no 4 (1979), pp. 5–21; and also Mitchell Dean, *Governmentality: Power and Rule in Modern Society* (London: Sage, 1999). Drawing upon both Weber and Foucault, Anthony Giddens lays particular stress on the 'administrative power' of the modern state; see his *A Contemporary Critique of Historical Materialism*, vol. II: *The Nation-State and Violence* (Cambridge: Polity, 1985), especially ch. 7.

46 Michel Foucault, *Power/Knowledge: Selected Interviews and Other Writings 1972–1977* (Brighton: Harvester, 1980).

47 Michel Foucault, *Politics, Philosophy, Culture: Interviews and Other Writings, 1977–1984*, ed. Lawrence Kritzman (London: Routledge, 1988), pp. 154–5.

Notes on Chapter 1

1 See for instance UNICEF, 'Water and sanitation briefing' (June 2001); ICRC, 'Iraq: a decade of sanctions' (14 December 1999); and on the reasons why the water situation in Iraq became so critical following the Gulf War, Barton Gellman, 'Allied air war struck broadly in Iraq; officials acknowledge strategy went beyond purely military targets', *Washington Post* (23 June 1991).

2 See for example George Monbiot, 'They're all damned', *Guardian* (26 February 2002); and articles on the Ilisu Dam Campaign webpage at http://www.ilisu.org.uk/.

3 John Vidal, 'Water of strife', *Guardian* (27 March 2002).

4 See for instance Oxfam, 'Crisis in southern Africa', briefing paper 23 (June 2002).

5 On Ghana see for example Christian Aid, 'Master or servant: how global trade can work for the benefit of poor people' (2001); and Integrated Social Development Centre, 'Water privatisation in Ghana: an analysis of

government and World Bank policies' (Accra: ISDC, 2001). On Bolivia see for example Jim Shutz, 'Globalization and the War for Water in Bolivia' (Cochabamba: Democracy Center, 2000); and for a very different analysis, Geraldine Dalton, 'Private Sector Finance for Water Infrastructure: What Does Cochabamba Tell Us About Using This Instrument?', SOAS Water Issues Group, Occasional Paper No 37 (2001). On South Africa see for instance Glenda Daniels, 'Water privatisation test case "a total debacle"', *Mail and Guardian* (Johannesburg) (16 November 2001). See also, for an important overview and critique, Maude Barlow and Tony Clarke, *Blue Gold: The Battle Against Corporate Theft of the World's Water* (London: Earthscan, 2002).

6 Max Weber, *The Methodology of the Social Sciences* Glencoe: Free Press (1949) p. 90.

7 Michel Foucault, *Power/Knowledge*, p. 133.

8 Malin Falkenmark and Jan Lundqvist, 'Looming water crisis: new approaches are inevitable', in Leif Ohlsson (ed.), *Hydropolitics: Conflicts Over Water as a Development Constraint* (London: Zed, 1995), p. 183.

9 Malin Falkenmark, 'Fresh water: time for a modified approach', *Ambio*, vol. 15, no 4 (1986) p. 192.

10 Worldwatch Institute, 'Populations outrunning water supply', press release (23 September 1999); Falkenmark and Lundqvist, 'Towards water security: political determination and human adaptation crucial', *Natural Resources Forum*, vol. 21, no 1 (1998) p. 41. It is worth mentioning that, despite its continued use of Malthusian ecological rhetoric, this more recent article is noticeably less ecological and indeed much more technical than Falkenmark and Lundqvist's earlier work.

11 Joyce Starr, 'Water wars', *Foreign Policy*, no 82 (1991) p. 17; Helena Lindholm, 'Water and the Arab–Israeli conflict', in Ohlsson, *Hydropolitics*, p. 57.

12 Malin Falkenmark, 'The massive water scarcity now threatening Africa: why isn't it being addressed?' *Ambio*, vol. 18, no 2 (1989), pp. 112–8; Falkenmark, 'Middle East hydropolitics: water scarcity and conflicts in the Middle East', *Ambio*, vol. 18, no 6 (1989), pp. 350–2; Falkenmark, 'Vulnerability generated by water scarcity', *Ambio*, vol. 18, no 6 (1989), pp. 352–3.

13 These figures are taken from Lowi, *Water and Power*, p. 29.

14 GTZ, 'Middle East regional study on water supply and demand development, concluding report', p. 6.

15 Ibid. If this figure does not appear to make sense given the Jordan River's estimated discharge of 1.6 bcmy, this is because that figure includes water flowing from Lebanon and Syria, whereas the other figures cited here denote natural resources available to Israel, Jordan and the West Bank and Gaza only, and do not include either Syria or Lebanon.

16 Israeli Central Bureau of Statistics, 'Monthly Bulletin of Statistics', vol. 51 (March 2000).

17 Figures from (and extrapolated from) Palestinian Central Bureau of

Statistics, 'The Palestinian Census of Population, Housing and Establishments: Preliminary Results' (December 1997).

18 Jordanian National Information System, Department of Statistics, 'Population projection 1998–2005' (1998).

19 On climate change and water crisis see for example Peter Gleick, 'The implications of global climatic changes for international security', *Climatic Change*, vol. 15 (1989) pp. 309–25.

20 On water as a constraint on development see for example Ohlsson, *Hydropolitics*.

21 See for instance Bulloch and Darwish, *Water Wars*; John Cooley, 'The war over water', *Foreign Policy*, no 54 (1984) pp. 3–26; Peter Gleick, 'Water, war and peace in the Middle East', *Environment*, vol. 36, no 3 (1994), pp. 6–15, 35–42; Starr, 'Water wars'. The question of 'water wars' will be addressed in greater depth in the next chapter.

22 Thomas Naff, in testimony, 26 June 1990; cited in Isam Shawwa, 'The water situation in the Gaza Strip' in Gershon Baskin (ed.), *Water: Conflict or Co-operation?* (Jerusalem: Israel/Palestine Center for Research and Information, 1992), p. 36.

23 Falkenmark and Carl Widstrand, 'Population and Water Resources: A Delicate Balance', *Population Bulletin*, vol. 47, no 3 (1992) p. 28; Worldwatch Institute, 'Populations Outrunning Water Supply'.

24 Quoted in 'Flowing uphill', *Economist* (12 August 1995) p. 46.

25 Interview with Saul Arlosoroff, Zafrir Weinstein Engineers and Consultants (5 April 1998).

26 For a general treatment see J. T. Winpenny, *Managing Water as an Economic Resource* (London: Routledge, 1994); and in relation to the Israeli–Palestinian arena, various articles in Marwan Haddad and Eran Feitelson, *Joint Management of Shared Aquifers: The Second Workshop* (Jerusalem: Palestine Consultancy Group and Harry S. Truman Institute for the Advancement of Peace, 1997), part 3.

27 McNamara, 'On gaps and bridges,' in *The Essence of Security: Reflections in Office* (London: Hodder and Stoughton, 1968), pp. 107–21.

28 Ohlsson, *Environment, Scarcity and Conflict: A Study of Malthusian Concerns*, Department of Peace and Conflict Research (Goteburg University 1999); and Anthony Turton, 'Water Scarcity and Social Adaptive Capacity: Towards an Understanding of the Social Dynamics of Water Demand Management in Developing Countries', SOAS Water Issues Group, Occasional Paper no 9 (1999).

29 Wolf, *Hydropolitics Along the Jordan River*, chs 3 and 4; Wolf and Masahiro Murakami, 'Techno-political decision making for water resource development: the Jordan River watershed', Water Resources Development, vol. 11, no 2 (1995) pp. 147–62.

30 World Bank, *From Scarcity to Security: Averting a Water Crisis in the Middle East and North Africa* (Washington DC: World Bank, 1995), p. 1.

31 Ibid.

32 World Bank, *Developing the Occupied Territories: An Investment in Peace*, vol. 5: Infrastructure (Washington DC: World Bank, 1993), p. 47.

33 World Bank, *Emergency Assistance Program for the Occupied Territories*, p. 8.

34 Sen, *Poverty and Famines: An Essay on Entitlement and Deprivation* (Oxford: Clarendon, 1981), p. 1.

35 Amongst the key dependency texts are Paul Baran, *The Political Economy of Growth* (New York: Monthly Review, 1957); Andre Gunther Frank, *Capitalism and Underdevelopment in Latin America* (London: Monthly Review, 1967); and Samir Amin, *Unequal Development* (Sussex: Harvester, 1976).

36 Jad Isaac, 'Core issues of the Palestinian–Israeli water dispute', in K. Spillman and G. Bachler (eds), *Environmental Crisis: Regional Conflicts and Ways of Cooperation, Environment and Conflicts Project*, Occasional Paper no 14 (Zurich: Centre for Security Studies and Conflict Research, 1995), p. 57.

37 These figures are extrapolated from Section 10 to Annex III of the Oslo II Agreement, and are widely used, especially by Palestinian commentators. The figures exclude the 78 mcmy which, as of 1995, still remained to be developed from the West Bank's aquifers.

38 GTZ, 'Middle East Regional Study on Water Supply and Demand Management', Concluding Report, p. 7.

39 On international water law see for instance Dante Caponera, *Principles of Water Law and Administration, National and International* (Rotterdam: Balkimi, 1992); and Stephen McCaffrey, 'Water, Politics and International Law', in Peter Gleick, *Water in Crisis: A Guide to the World's Fresh Water Resources* (New York: Oxford University Press, 1993), pp. 92–104.

40 Jeffrey Dillman, 'Water Rights in the Occupied Territories', *Journal of Palestine Studies*, vol. 19, no 1 (1989) pp. 46–71; Jamal El-Hindi, 'Note: the West Bank aquifer and conventions regarding laws of belligerent occupation', *Michigan Journal of International Law*, vol. 11, no 4 (1990) pp. 1400–23.

41 Hisham Zarour and Jad Isaac, 'Nature's apportionment and the open market: a promising solution to the Arab–Israeli water conflict', *Water International*, vol. 18, no 1 (1993) pp. 40–53.

42 Shuval, 'Approaches to resolving the water conflicts between Israel and her neighbours – a regional water-for-peace plan', *Water International*, vol. 17, no 3 (1992) pp. 133–43; 'Approaches to finding an equitable solution to the water resources problems shared by Israelis and the Palestinians in the use of the mountain aquifer', in Baskin, *Water: Conflict or Cooperation?*, pp. 37–84; and 'Proposed principles and methodology for the equitable allocation of the water resources shared by the Israelis, Palestinians, Jordanians, Lebanese and Syrians', in Assaf et al., *A Proposal*

for the Development of a Regional Water Master Plan (Jerusalem: Israel/Palestine Centre for Research and Information, 1993), pp. 150–74.

43 See for example James Moore, 'Parting the waters: calculating Israeli and Palestinian entitlements to the West Bank aquifers and the Jordan River basin', *Middle East Policy*, vol. 3, no 1 (1993) pp. 91–108; and Moore, 'An Israeli–Palestinian water sharing regime', in Isaac and Shuval, *Water and Peace in the Middle East*, pp. 181–92.

44 A phrase taken from Neil Smith, *Uneven Development: Nature, Capital and the Production of Space*, 2nd edn (Oxford: Blackwell, 1990), ch. 2.

45 Thomas Malthus, *An Essay on the Principle of Population* (Cambridge: Cambridge University Press, 1989).

46 Donella Meadows et al., *The Limits to Growth: A Report for the Club of Rome's Project on the Predicament of Mankind* (New York: Earth Island, 1972); Paul Ehrlich, *The Population Bomb* (New York: Ballantine, 1968); Paul Ehrlich and Anne Ehrlich, *The Population Explosion* (New York: Touchstone, 1991); Jonathan Porrit, *Seeing Green: The Politics of Ecology Explained* (Oxford: Basil Blackwell, 1984), p. 49.

47 Quoted in S. Peterson, 'What could float – or sink – peacemaking', *Christian Science Monitor* (14 July 1999).

48 Falkenmark and Lundqvist, 'Looming water crisis', p. 183.

49 Wolf, *Hydropolitics Along the Jordan River*, p. 7.

50 Ibid., pp. 91–3.

51 Marx, *Early Writings*, ed. T.B. Bottomore (London: Watts and Co, 1963), p. 161.

52 Marx and Friedrich Engels, *The German Ideology* (London: Lawrence and Wishart, 1965), p. 59.

53 Bill McKibben, *The End of Nature* (New York: Viking, 1990).

54 Wolf, *Hydropolitics Along the Jordan River*, p. 7; additional figures from Arnon Soffer, 'The relevance of the Johnston Plan to the reality of 1993 and beyond', in Isaac and Shuval (eds), *Water and Peace in the Middle East*, p. 114.

55 T. Ben-Gai et al., 'Long-term change in October rainfall patterns in southern Israel', *Theoretical and Applied Climatology*, vol. 46, no 4 (1993) pp. 209–17; Ben-Gai et al., 'Long-term changes in annual rainfall patterns in southern Israel', *Theoretical and Applied Climatology*, vol. 49, no 2 (1994) pp. 59–67.

56 On the Western imaginary, see for example Kate Soper, *What is Nature?* (Oxford: Blackwell, 1995); and for a rather more bizarre analysis, Bruno Latour, *We Have Never Been Modern* (Cambridge, MA: Harvard University Press, 1993).

57 Such is the situation on Frank Herbert's mythical desert planet of Arakis where, in the absence of plentiful water supplies, the local population survives by wearing 'stillsuits' which minimise water losses and recycle sweat. Without wanting to suggest that Middle Easterners should start

wearing stillsuits, Herbert's *Dune* (London: Hodder and Stoughton, 1965) is nevertheless illustrative, in the extreme, of the real nature of technology-resource relations.

58 Marx, *Capital.* vol. I, London: Pelican (1976) p. 283.

59 Harvey, *Justice, Nature and the Geography of Difference* (Oxford: Blackwell, 1996), p. 147.

60 For discussion of contrasting 'fluid' and 'territorial' representations of space, see for instance Gilles Deleuze and Felix Guattari, *A Thousand Plateaus: Capitalism and Schizophrenia* (London: Athlone, 1987), ch. 1 (which distinguishes between 'rhizomatic' and 'arboreal' representations); and Michel Serres and Bruno Latour, *Conversations on Science, Culture and Time* (Ann Arbor: University of Michigan Press, 1995), especially pp. 107–12 (on 'circulations' and 'reservoirs'). For discussion of these issues in relation to water see Eric Swyngedouw, 'Hybrid waters: on water, nature and society', presented at a conference on 'Sustainability, Risk and Nature: The Political Ecology of Water in Advanced Societies', University of Oxford (15–17 April 1999).

61 J. Anthony Jones, *Global Hydrology: Processes, Resources and Environment* (Harlow: Longman, 1997), p. 320.

62 A. Kanarek and M. Michail, 'Groundwater recharge with municipal effluent: Dan Region Reclamation Project, Israel', *Water Science and Technology*, vol. 34, no 11 (1996) pp. 227–33.

63 Allan, 'Water in the Middle East and in Israel–Palestine: some local and global resource issues', in Haddad and Feitelson, *Joint Management of Shared Aquifers: The Second Workshop*, pp. 31–44.

64 Allan, 'Global Systems Ameliorate Local Droughts: Water, Food and Trade', *SOAS Water Issues Group*, Occasional Paper no 10 (1999). For fuller discussion see Allan, *The Middle East Water Question.*

65 'Government to create water desalting facility', *Ha'aretz* (18 April 2000); Aluf Benn, 'Import water from Turkey, prime minister's advisor urges', *Ha'aretz* (30 March 2000); Amiram Cohen, 'Construction begins on imported water pipe', *Ha'aretz* (30 August 2000).

66 On the medusa bag scheme see for instance Hugh Pope, 'Water in a bag', *Middle East International*, no 377 (8 June 1990) p. 14; on the 'peace canal', Boaz Wachtel, 'The peace canal project: a multiple conflict resolution perspective for the Middle East', in Isaac and Shuval, *Water and Peace in the Middle East*, pp. 363–73; and on the Red-Dead and Med-Dead Canals, Murakami, Managing Water for Peace in the Middle East: Alternative Strategies (Tokyo: United Nations University Press, 1995), ch. 5.

67 Peter Beaumont, 'Conflict, coexistence, and cooperation: a study of water use in the Jordan basin', in Amery and Wolf (eds), *Water in the Middle East*, p. 33.

68 Ibid.

69 Harvey, *Justice, Nature and the Geography of Difference*, p. 147.

70 See Harvey's excellent discussion in *Justice, Nature and the Geography of Difference*, pp. 139–49.
71 Seregaldin, quoted in 'Flowing uphill', p. 46.
72 I allude here to the report of the Brundtland Commission: *World Commission on Environment and Development, Our Common Future* (Oxford: Oxford University Press, 1987).
73 For a classic statement of such 'modernisation' development thinking see Walter Rostow, *The Stages of Economic Growth: A Non-Communist Manifesto* (Cambridge: Cambridge University Press, 1960); and for a paradigmatically 'internalist' liberal thesis, this one on the relationship between a state's internal make-up and its international behaviour, see Michael Doyle, 'Kant, liberal legacies, and foreign affairs: parts 1 and 2', *Philosophy and Public Affairs*, vol. 12 (1983), pp. 205–34 and 323–53, critiqued as internalist (or what he calls 'reductionist') by Kenneth Waltz, in Fred Halliday and Justin Rosenberg, 'Interview with Ken Waltz', *Review of International Studies*, vol. 24 (1998) pp. 371–86.
74 World Bank, *Developing the Occupied Territories*, p. 7.
75 The notion of 'dependent development' is developed by Fernando Cardoso and Enzo Faletto, *Dependency and Development in Latin America* (Berkeley: University of California Press, 1979). The Palestinian economy is characterised as 'dependent' by Yezid Sayigh, 'The Palestinian economy under occupation: dependency and pauperization', *Journal of Palestine Studies*, vol. 15, no 4 (1986), pp. 46–67; while the phrase 'de-development' is developed by Sara Roy, *The Gaza Strip: The Political Economy of De-Development* (Washington DC: Institute for Palestine Studies, 1995).
76 James Ferguson argues along very similar lines with regard to the Lesotho economy in his brilliant book *The Anti-Politics Machine: 'Development', Depoliticization and Bureaucratic Power in Lesotho* (Cambridge: Cambridge University Press, 1990).
77 Isaac, 'Core issues of the Palestinian–Israeli water dispute', p. 59.
78 Foucault, *Power/Knowledge*, p. 183.
79 Foucault, *Politics, Philosophy, Culture*, p. 118.
80 Foucault, *Power/Knowledge*, pp. 92, 96.
81 Foucault, 'Afterword: the subject and power', in Hubert Dreyfus and Paul Rabinow, *Michel Foucault: Beyond Structuralism and Hermeneutics* (Hemel Hempstead: Harvester Wheatsheaf, 1982), pp. 216–7.
82 Foucault, *Discipline and Punish* (London: Penguin, 1977), pp. 26, 139.
83 Winner, *The Whale and the Reactor: A Search for Limits in an Age of High Technology* (Chicago: Chicago University Press, 1986), ch. 1.
84 One striking exception here is the work of Julie Trottier, especially her *Hydropolitics in the West Bank and Gaza*. Trottier's work is clearly critical rather than problem-solving in orientation, and indeed has a great deal in common with the present study.

85 This having been most clearly the case in the multilateral arm of the peace process; see Joel Peters, *Pathways to Peace: The Multilateral Arab–Israeli Peace Talks* (London: Royal Institute for International Affairs, 1996), especially pp. 16–22.
86 For discussion see Bruce Rich, *Mortgaging the Earth: The World Bank, Environmental Impoverishment and the Crisis of Development* (London: Earthscan, 1994), pp. 57–8.
87 Wolf, *Hydropolitics Along the Jordan River*, p. 139.
88 Lowi, *Water and Power*, p. xxi.
89 Rouyer, *Turning Water into Politics*, p. 9.
90 Rich, *Mortgaging the Earth*, p. 76.

Notes on Chapter 2

1 Dolatyar and Gray, 'The politics of water scarcity in the Middle East', *Environmental Politics*, vol. 9, no 3 (2000) pp. 65, 84.
2 Starr and Stoll (eds), *The Politics of Scarcity: Water in the Middle East* (Boulder: Westview, 1988), p. ix.
3 Boutros Ghali in the *Independent on Sunday* (6 May 1990) as quoted in Ewan Anderson, 'The political and strategic significance of water', *Outlook on Agriculture* (December 1992); Serageldin in *Financial Times* (7 August 1995), as quoted in Dolatyar and Gray, *Water Politics in the Middle East*, pp. 8, 22.
4 Starr, *Covenant Over Middle East Waters*.
5 Boutros Ghali's ill-timed comments were made just three months before the Iraqi invasion of Kuwait.
6 Bulloch and Darwish, *Water Wars*, p. 34.
7 Thomas Naff and Ruth Matson (eds), *Water in the Middle East: Conflict or Cooperation?* (Boulder: Westview, 1984), p. 44.
8 Cooley, 'The war over water', p. 3.
9 The informed commentaries include Sidney Bailey, *Four Arab–Israeli Wars and the Peace Process* (London: Macmillan, 1990); Michael Brecher, *Decisions in Crisis: Israel, 1967 and 1973* (Berkeley: University of California Press, 1980); Trevor Dupuy, *Elusive Victory: The Arab–Israeli Wars, 1947–1974* (London: Macdonald and Jane's, 1978); Norman Finkelstein, *Image and Reality of the Israel–Palestine Conflict* (London: Verso, 1995), ch. 5; Fawaz Gerges, *The Superpowers and the Middle East: Regional and International Politics, 1955–1967* (Boulder: Westview, 1994); Walter Laqueur, *The Road to War: The Origins and Aftermath of the Arab–Israeli Conflict, 1967–8* (London: Weidenfeld and Nicolson, 1968); Donald Neff, *Warriors for Jerusalem: The Six Days that Changed the Middle East* (New York: Simon and Schuster, 1984); Richard Parker, *The Politics of Miscalculation in the Middle East* (Bloomington: Indiana University Press, 1993); and Nadav Safran, *From War to War: The Arab–Israeli Confrontation 1948–1967* (New

York: Pegasus, 1969). Without wanting to enter too far into the debates about the precise causes of the 1967 war, Avi Shlaim in my view gets it right in his observation that 'the war owed more to the rivalries between the Arab states than to the dispute between them and Israel'; see Shlaim, 'The Middle East: the origins of the Arab–Israeli wars', in Ngaire Woods (ed.), *Explaining International Relations Since 1945* (Oxford: Oxford University Press, 1996), p. 227. For all Israel's propaganda about the existential threats to its security, the reality was somewhat different. Take the words of Yitzhak Rabin, then Chief of Staff of the Israeli Defense Forces: 'I do not think Nasser wanted war. The two divisions he sent into the Sinai on May 14 would not have been sufficient to launch an offensive against Israel. He knew it and we knew it.'

10 Joyce Starr, 'Water wars', p. 19; Bulloch and Darwish, *Water Wars*, p. 198; Kliot, *Water Resources and Conflict in the Middle East*, p. 12.

11 Allan, *The Middle East Water Question*, p. 3.

12 I allude here to Edward Said's scathing critique of 'Western' scholarship and discourse on the Middle East and Islam, in particular *Orientalism: Western Conceptions of the Orient* (London: Penguin, 1978); and *Covering Islam: How the Media and the Experts Determine How We See the Rest of the World* (London: Routledge and Kegan Paul, 1981).

13 Deborah Gerner and Philip Schrodt, 'Middle East politics', in Gerner (ed.), *Understanding the Contemporary Middle East* (Boulder: Lynne Rienner, 2000), p. 90.

14 Dolatyar and Gray, 'The politics of water scarcity'; and *Water Politics in the Middle East.*

15 The canonical works being E.H. Carr, *The Twenty Years' Crisis, 1919–1939*, 2nd edn, (London: Macmillan, 1946); Hans Morgenthau, *Politics Among Nations: The Struggle for Power and Peace*, 6th edn (New York: Knopf, 1985); and Waltz, *Theory of International Politics* (Reading MA: Addison-Wesley, 1979).

16 Morgenthau, *Politics Among Nations*, p. 5.

17 Waltz, 'Realist thought and neorealist theory', *Journal of International Affairs*, vol. 44, no 1 (1990), p. 34.

18 The most well-known late twentieth-century iteration of this thesis is Michael Doyle, 'Kant, liberal legacies, and foreign affairs: parts 1 and 2', *Philosophy and Public Affairs*, vol. 12 (1983). Doyle looks back to Immanuel Kant's vision of 'Perpetual peace', collected in Kant's *Political Writings*, ed. H. Reiss (Cambridge: Cambridge University Press, 1970).

19 See for example Richard Rosencrance, *The Rise of the Trading State* (New York: Basic Books, 1986).

20 Joseph Grieco, *Cooperation Among Nations: Europe, America and Non-Tariff Barriers to Trade* (London: Cornell University Press, 1990), p. 4.

21 Ibid., p. 47. Robert Keohane develops hegemonic stability theory in 'The theory of hegemonic stability and changes in international economic

regimes, 1967–1977', in Ole Holsti et al. (eds), *Change in the International System* (Boulder: Westview, 1980), pp. 131–62. Keohane has since changed his tune somewhat, arguing along neo-liberal lines that co-operative arrangements first forged by hegemonic actors can remain powerful and effective even in the face of hegemonic decline. See Keohane, *After Hegemony: Co-operation and Discord in the World Political Economy* (Princeton: Princeton University Press, 1984).

22 On regimes, see for example Oran Young, *International Cooperation: Building Regimes for Natural Resources and the Environment* (Ithaca NY: Cornell University Press, 1989). The notion of 'epistemic communities' is developed by Peter Haas, *Saving the Mediterranean: The Politics of International Environmental Cooperation* (New York: Columbia University Press, 1990); and Haas (ed.) 'Knowledge, power and international policy coordination', *International Organization* (special issue), vol. 46, no 1 (1992).

23 See for example, David Mitrany, *A Working Peace System* (London: Royal Institute for International Affairs, 1943); Mitrany, *The Functional Theory of Politics* (New York: St. Martin's Press, 1975); and James Sewell, *Functionalism and World Politics* (Princeton: Princeton University Press, 1966).

24 Dolatyar and Gray, *Water Politics in the Middle East*, p. 209.

25 Dolatyar and Gray, 'The politics of water scarcity', p. 67.

26 Ibid., pp. 65, 84.

27 Dolatyar and Gray, *Water Politics in the Middle East*, p. 209.

28 Hillel, *Rivers of Eden*, p. 283; quoted Dolatyar and Gray, 'The politics of water scarcity,' p. 71; Wolf, *Hydropolitics Along the Jordan River*, p. 3; quoted Dolatyar and Gray, *Water Politics in the Middle East*, p. 113 (my italics).

29 Dolatyar and Gray, 'The politics of water scarcity', p. 83. On Saudi–Bahraini relations, see for example Hassan Hamdan Al-Alkim, *The GCC States in an Unstable World: Foreign Policy Dilemmas of Small States* (London: Saqi, 1994); and Anthony Cordesman, *Bahrain, Oman, Qatar and the UAE: Challenges of Security* (Boulder: Westview, 1997).

30 Undala Alam, *Water Rationality: Mediating the Indus Treaty*, PhD Thesis (University of Durham, 1998), pp. 263–4.

31 Dolatyar and Gray, *Water Politics in the Middle East*, p. 212.

32 Dolatyar and Gray 'The politics of water scarcity', pp. 67, 70.

33 Lowi, *Water and Power*, p. 196.

34 Ibid., p. xv.

35 Ibid., p. 10.

36 Ibid., p. 203.

37 Ibid., p. 10.

38 Ibid., pp. 118, 197.

39 Ibid., pp. 164–6.

40 Ibid., pp. 163–4.

41 Shlaim, *The Politics of Partition: King Abdullah, The Zionists and Palestine, 1921–1951* (Oxford: Oxford University Press, 1990), p. 2.
42 Simha Flapan, *The Birth of Israel: Myths and Realities* (London: Croom Helm, 1987), ch. 4.
43 Moshe Zak, 'Israeli–Jordanian negotiations', *Washington Quarterly*, vol. 8, no 1 (1985) pp. 167–76; and Adam Garfinkle, *Israel and Jordan in the Shadow of War: Functional Ties and Futile Diplomacy in a Small Place* (London: Macmillan, 1992).
44 Zak, 'Israel–Jordanian negotiations'.
45 Interview with Moshe Yizraeli, Consultant to the Israeli Water Commission (21 July 1998); Wolf, *Hydropolitics Along the Jordan River*, p. 48.
46 Shapland, *Rivers of Discord*, p. 16.
47 Lowi, *Water and Power*, p. 165.
48 Ibid., p. 8. For such a rational choice approach as applied to the Jordan basin, see M. Hirsch, 'Game theory, international law and future environmental cooperation in the Middle East', *Denver Journal of International Law and Policy*, vol. 27 (1998) pp. 75–119.
49 Ibid., p. 198.
50 Robert Gilpin, *War and Change in World Politics* (New York: Cambridge University Press, 1981), p. 227.
51 Cox, 'Social forces, states and world orders', p. 131.
52 Falkenmark and Lundqvist, 'Looming water crisis', p. 183.
53 Lowi, *Water and Power*, p. 202.
54 Ibid., p. xv.
55 Frederick Frey and Thomas Naff, 'Water: an emerging issue in the Middle East?' *Annals of the American Academy of Political and Social Science*, vol. 482 (1985) p. 67.
56 Interview with Saul Arlosoroff.
57 Alam, *Water Rationality*; Dolatyar and Gray, 'The politics of water scarcity', p. 67.
58 Gershon Shafir, *Land, Labour and the Origins of the Israeli–Palestinian Conflict, 1882–1914* (Cambridge: Cambridge University Press, 1989).
59 Shafir and Yoav Peled, *Being Israeli: The Dynamics of Multiple Citizenship* (Cambridge: Cambridge University Press, 2002), p. 37. Shafir, it should be said, is not alone in having developed such a colonisation perspective. The argument is common within Palestinian thought – see for instance Elia Zureik, *The Palestinians in Israel: A Study in Internal Colonialism* (London: Routledge and Kegan Paul, 1978) – and elsewhere, the theme was famously formulated by Maxime Rodinson in *Israel: A Colonial-Settler State?* (New York: Monad, 1973). Within Israeli sociology, the perspective is associated not just with Shafir but also with the work of Baruch Kimmerling, especially his *Zionism and Territory: The Socio-Territorial Dimension of Zionist Politics* (Berkeley: University of California Press, 1983), where he develops a Weberian account of the early Zionist settlement. For reasons

that do not need detailing here, I find Shafir's Marxist account the more persuasive of the two.

60 Shafir and Peled, *Being Israeli*, p. 38.

61 Shalev, 'The political economy of Labor Party dominance and decline in Israel', in T.J. Pempel (ed.), *Uncommon Democracies: The One-Party Dominant Regimes* (Ithaca: Cornell University Press, 1990), pp. 83–127.

62 See especially Talcott Parsons, *The Structure of Social Action* (New York: Free Press, 1949); Shmuel Eisenstadt, *Israeli Society* (London: Weidenfeld and Nicolson, 1967).

63 Kimmerling, 'Boundaries and frontiers of the Israeli control system', in Kimmerling (ed.), *The Israeli State and Society: Boundaries and Frontiers* (Albany: State University of New York Press, 1989), p. 270.

64 Shafir, *Land, Labour and the Origins of the Israeli–Palestinian Conflict*, p. xii.

65 Itzhak Galnoor, 'Water Planning: Who Gets the Last Drop?' in R. Bilski (ed.), *Can Planning Replace Politics? The Israeli Experience* (The Hague: Martinus Nijhoff 1980), pp. 137–215; Galnoor, 'Water Policymaking in Israel', in Hillel Shuval (ed.), *Water Quality Management Under Conditions of Scarcity: Israel as a Case Study* (New York: Academic Press 1980) pp. 287–314; Rouyer, *Turning Water into Politics*, ch. 3; and Rouyer, 'Zionism and water: influences on Israel's future water policy during the pre-state period', *Arab Studies Quarterly*, vol. 18, no 4 (1996), pp. 25–47. Lowi (*Water and Power*, pp. 51–2) and Dolatyar and Gray ('The politics of water scarcity', p. 69) also both allude to the ideological importance of water within Israel, but operating as they do with a faith in the transhistorical rationality of state action, nonetheless fail to recognise its historical and political significance.

66 See for instance Shlomo Avineri, *The Making of Modern Zionism: The Intellectual Origins of the Jewish State* (London: Weidenfeld and Nicolson, 1981), ch. 14, on the philosophy of A.D. Gordon.

67 Rouyer, *Turning Water into Politics*, p. 93; J. Hurewitz, *Diplomacy in the Near and Middle East: A Documentary Record: 1914–1956*, vol. 2 (New York: Octagon, 1956), p. 28.

68 Elmusa, *Water Conflict*, p. 277.

69 Rouyer, *Turning Water into Politics*, pp. 154–6.

70 Ibid., pp. 8–9.

71 Interview, 5 April 1998.

72 Ibid.

73 Ibid.

74 Cited in Galnoor, 'Water Planning', p. 159.

75 Feitelson, 'Implications of shifts in the Israeli water discourse for Israeli–Palestinian water negotiations', *Political Geography*, vol. 21 (2002), p. 305.

76 Ibid., p. 300.

77 Feitelson, 'The ebb and flow of the Arab–Israeli water conflict: are past confrontations likely to resurface?' *Water Policy*, vol. 2 (2000), p. 357.

78 Rouyer, *Turning Water into Politics*, p. 26.
79 Gershon Baskin, 'The West Bank and Israel's water crisis', in Baskin, *Water*, p. 9.
80 Ministry of Agriculture and Rural Development, 'Agricultural economic report for 1996' (1997) (in Hebrew); quoted in Amnon Kartin, 'Factors inhibiting structural changes in Israel's water policy', *Political Geography*, vol. 19 (2000), p. 109.
81 Ben Meir, 'The water issue in Israel', Conference on Water for Israel: Resources, Utilizations and Problems (April 1997) (in Hebrew); quoted in Kartin, 'Factors inhibiting structural changes in Israel's water policy', p. 108.
82 Migdal, *Strong Societies and Weak States: State–Society Relations and State Capabilities in the Third World* (Princeton: Princeton University Press, 1988); Migdal, *Through the Lens of Israel: Explorations in State and Society* (Albany: State University of New York Press, 2001).
83 Ephraim Kleiman, 'The place of manufacturing in the growth of the Israeli economy', *Journal of Development Studies*, vol. 3, no 3 (1967), p. 233.
84 Shafir and Peled, *Being Israeli*, p. 58.
85 Ben-Eliezer, *The Making of Israeli Militarism* (Bloomington: Indiana University Press, 1998); Ben-Eliezer, 'Is a military coup possible in Israel? Israel and French-Algeria in comparative historical-sociological perspective', *Theory and Society*, vol. 27 (1998), pp. 311–49.
86 Discussed for instance in Arlosoroff, 'The Israeli water law concept: summary', in Feitelson and Haddad, *Joint Management of Shared Aquifers: The Fourth Workshop*, pp. 111–5.
87 On these see for example Elmusa, *Water Conflict*, pp. 261–2; Galnoor, 'Water planning', pp. 147–8; Lonergan and Brooks, *Watershed*, pp. 59–61; and Rouyer, *Turning Water into Politics*, pp. 148–52.
88 Galnoor, 'Water planning', p. 172.
89 Kenneth Stein, *The Land Question in Palestine, 1917–1939* (London: Chapel Hill, 1984).
90 See on this much discussed and controversial topic Nur Masalha, *Expulsion of the Palestinians: The Concept of Transfer in Zionist Political Thought, 1882–1948* (Washington DC: Institute for Palestine Studies, 1992); Benny Morris, *The Birth of the Palestinian Refugee Problem, 1947–1949* (Cambridge: Cambridge University Press, 1987); Nur Masalha's critique of Morris's arguments therein, 'A critique on Benny Morris', *Journal of Palestine Studies*, vol. 21, no 1 (1991), pp. 90–7; Finkelstein's critique of Morris in *Image and Reality of the Israel–Palestine Conflict*, ch. 3; and Ilan Pappe, *The Making of the Arab–Israeli Conflict, 1947–1951* (London: I.B.Tauris, 1992).
91 Shafir and Peled, *Being Israeli*, p. 111.
92 See for instance Ian Lustick, *Arabs in the Jewish State: Israel's Control of a National Minority* (Austin: University of Texas Press, 1980).

93 Shafir and Peled, *Being Israeli*, p. 117.
94 Aziz Haidar, *On the Margins: The Arab Population in the Israeli Economy* (London: Hurst and Co, 1995), p. 48.
95 David McDowall, *The Palestinians: The Road to Nationhood* (London: Minority Rights Publications, 1994), p. 55.
96 Uri Davis, *Israel: An Apartheid State* (London: Zed, 1987).
97 On the Arab League diversion and the Israeli response see Lowi, *Water and Power*, ch. 5; and on Israeli–Syrian negotiations over the Golan Heights, and the importance of water issues therein, Michael Jansen, 'The peace process flounders in Geneva', *Middle East International* (7 April 2000).

Notes on Chapter 3

1 Quoted in Neff, *Warriors for Jerusalem*, p. 299.
2 Kimmerling, *Zionism and Territory*, pp. 157–66.
3 Shafir and Peled, *Being Israeli*, p. 161.
4 Ibid.
5 David Newman, *The Impact of Gush Enumin: Politics and Settlement in the West Bank* (London: Croom Helm, 1985).
6 Yediot Achronot (18 June 1976); quoted in Shafir and Peled, *Being Israeli*, p. 168.
7 Meron Benvenisti, *The West Bank Data Project: A Survey of Israel's Policies* (Washington DC: American Enterprise Institute, 1984), ch. 6.
8 Shafir and Peled, *Being Israeli*, pp. 172–3.
9 Benvenisti, *The West Bank Data Project*, ch. 6.
10 Foundation for Middle East Peace, *Report on Israeli Settlement*, vol. 3, no 5 (1993).
11 See especially George Abed (ed.), *The Palestinian Economy* (London: Routledge, 1988); Benvenisti, *The West Bank Data Project*, ch. 2; Roy, *The Gaza Strip*; Cheryl Rubenberg, 'Twenty years of Israeli economic policies in the West Bank and Gaza: prologue to the intifada', *Journal of Arab Affairs*, vol. 8, no 1 (1989), pp. 28–73; Samir Abdallah Saleh, 'The effects of Israeli occupation on the economy of the West Bank and Gaza Strip', in Jamal Nassar and Roger Heacock (eds), *Intifada: Palestine at the Crossroads* (New York: Praeger, 1991), pp. 37–51; Adel Samara, *The Political Economy of the West Bank 1967–1982* (London: Khamsin, 1989); and Sayigh, 'The Palestinian economy under occupation'.
12 Adel Samara in Efrat, 'Interview with Adel Samara: the hidden logic of Arafat's surrender', *Challenge* (June 2000). The true figure is uncertain, with estimates ranging up to as many as 189,000 (Usher, *Palestine in Crisis*, p. 6).
13 Sayigh, 'The Palestinian economy under occupation', p. 47.

14 The best account of this regime is Eyal Benvenisti, *Legal Dualism: The Absorption of the Occupied Territories into Israel* (Boulder: Westview, 1990).

15 Military Proclamation No 2 (7 June 1967); quoted in Elmusa, *Water Conflict*, p. 263.

16 Meron Benvenisti, *Report: Demographic, Economic, Legal, Social and Political Developments in the West Bank* (Boulder: Westview, 1986), p. 56; discussed in Shafir and Peled, *Being Israeli*, p. 176.

17 On this in relation to the West Bank see especially Emile Sahliyeh, *In Search of Leadership: West Bank Politics since 1967* (Washington DC: Brookings Institution, 1988); and Don Peretz, *The West Bank: History, Politics, Society and Economy* (Boulder: Westview, 1986).

18 *New York Times* (28 May 1985); quoted in Sharif and Peled, *Being Israeli*, p. 194.

19 Chomsky, *Fateful Triangle*, p. 129; Ibrahim Abu-Lughod, 'Introduction: on achieving independence', in Nassar and Heacock, *Intifada*, p. 10. On the intifada more generally see for instance Chomsky, *Fateful Triangle*, ch. 8; Zachary Lockman and Joel Beinin (eds), *Intifada: The Palestinian Uprising Against Israeli Occupation* (Washington DC: MERIP, 1989); Andrew Rigby, *Living the Intifada* (London: Zed, 1991); and Ze'ev Schiff and Ehud Ya'ari, *Intifada: The Palestinian Uprising – Israel's Third Front* (New York: Simon and Schuster, 1989).

20 Menahim Cantor, quoted in Samara, *The Political Economy of the West Bank*, p. 80. On these water policies see especially Uri Davis, Antonia Maks and John Richardson, 'Israel's water policies', *Journal of Palestine Studies*, vol. 9, no 2 (1980), pp. 3–31; Elmusa, *Water Conflict*, chs. 2 and 4; Jerusalem Media and Communication Centre, *Water: The Red Line* (Jerusalem: JMCC, 1994), ch. 4; Rouyer, *Turning Water into Politics*, ch. 2; Gwyn Rowley, 'The West Bank: native water resource systems and competition', *Political Geography Quarterly*, vol. 9, no 1 (1990), pp. 39–52; and Joe Stork, 'Water and Israel's occupation strategy', *Middle East Report*, vol. 13, no 6 (1983), pp. 19–24.

21 Military Order no 92 (15 August 1967), as cited in Elmusa, *Water Conflict*, p. 265.

22 Elmusa, *Water Conflict*, pp. 265–6; Rouyer, *Turning Water into Politics*, p. 48; Raja Shehadeh, *Occupier's Law: Israel and the West Bank* (Washington DC: Institute for Palestine Studies, 1988).

23 As quoted in Elmusa, *Water Conflict*, p. 266.

24 Ibid., p. 50; interview with Taher Nassereddin, Palestinian Head, West Bank Water Department (12 April 1998).

25 Rouyer, *Turning Water into Politics*, p. 48.

26 Lonergan and Brooks, *Watershed*, p. 130.

27 State Comptroller of Israel, 'Report on the Management of Water Resources in Israel' (Jerusalem: Government of Israel, 1990). For discussion, see Rouyer, *Turning Water into Politics*, p. 53; and Jerusalem Media and Communication Centre, *Water*, p. 46.

28 Elmusa, *Water Conflict*, pp. 270–1.
29 Interview with Taher Nassereddin (12 April 1998).
30 Elmusa, *Water Conflict*, pp. 271–2; and interview with Taher Nassereddin (12 April 1998).
31 Interview with Taher Nassereddin (12 April 1998).
32 Ibid., and interview with Abdul Rahman Tamimi, General Director, Palestinian Hydrology Group (19 April 1998).
33 Isaac and Jan Selby, 'The Palestinian water crisis: status, projections and potential for resolution', *Natural Resources Forum*, vol. 20 (1996), pp. 18–20.
34 Shlomo Gazit, *The Carrot and the Stick: Israeli Command of the West Bank and Gaza* (Nicosia: Beisan, 1985), pp. 91–2.
35 David Scarpa, 'The southern West Bank aquifer: exploitation and sustainability', in J. Ginal and J. Ragep (eds), *Water in the Jordan Valley: Technical Solutions and Regional Cooperation* (Oklahoma: University of Oklahoma Press, 2002).
36 *Turning Water into Politics*, p. 48.
37 The following account is based largely on CDM/Morganti, 'Task 25: Water Supply Facility Master Plan for the Hebron-Bethlehem Service Area, Final Report', *Report for the PWA* (8 March 1997); and PWA, 'West Bank Water Facilities Map' (1996). Use has also been made of CDM/Morganti, 'Task 4: Comprehensive Planning Framework for Palestinian Water Resources Development, Final Report, vol. 2', *Report for the PWA* (6 July 1997); CDM/Morganti, 'Task 33: Integrated Water Resources Management Plan for the Hebron-Bethlehem Area, Final Plan', *Report for the PWA* (5 September 1997), especially pp. 26–32; ARIJ, 'Environmental Profile for the West Bank, vol. 1: District of Bethlehem' (1995); 'Environmental Profile for the West Bank, vol. 3: Hebron District' (1995); and interview with Taher Nassereddin, West Bank Water Department (15 August 1998).
38 CDM/Morganti, 'Task 25', p. 6 puts 1995 pumpage from ten of these twelve wells at 12.9 mcmy; the other two wells, located between Beit Sahour and Jerusalem, produced 0.5 mcmy in 1996, and I assume roughly the same figure for 1995 (PWA, 'West Bank Water Facilities Map').
39 CDM/Morganti, 'Task 25', p. 9.
40 Ibid.; and also interview with Taher Nassereddin (15 August 1998).
41 Ibid., p. 13.
42 CDM/Morganti, 'Task 25', p. 7.
43 Ibid., p. 13.
44 Ibid., pp. 11–12.
45 Ibid., pp. 7–12.
46 Population figures are taken, for Palestinian communities, from CDM/Morganti, 'Task 33', Appendices 2–5 (these figures are from the Palestinian Central Bureau of Statistics); and for Israeli settlements,

Foundation for Middle East Peace, 'Settlements in the West Bank'. Figures from both sources are for 1996, and have not been corrected for 1995.

47 This figure for Dhahriyya is taken from CDM/Morganti, 'Task 33', p. 94.
48 CDM/Morganti, 'Task 25', p. 9.
49 Ibid., p. 11.
50 Ibid., pp. 11–12.
51 Ibid., p. 7.
52 This account is based on Andrea Cippa, 'Water Management in the City of Hebron', TIPH Information Report no 1945 (Hebron, October 1997), pp. 9–10; SOGREAH Ingenierie, 'Hebron Municipality Hydraulic Sketch' (1996); interview with Ziad Qannam, WSERU (30 March 1998); and meeting with Andre Naudy, Vivendi and General Des Eaux, Halhoul (25 August 1999).
53 My knowledge of the situation in Duwarra is drawn from informal conversations with residents during the course of three visits to the village (all summer 1998), and in particular with Idris Shaker Jabarin, Head, Village Council (various 1998). I should mention also informal conversations with Lucianne Maas, Samira Refa'ai and Walid Shalan, all Save the Children Federation, Halhoul (various occasions, 1998 and 1999); and interview with Taher Nassereddin (15 August 1998).
54 CDM/Morganti, 'Task 33', pp. 92–3, 96; World Bank, 'Developing the Occupied Territories, vol. 5: Infrastructure', p. 45.
55 SOGREAH Ingenierie, 'Master Plan for Water Distribution in the Bethlehem Area, Interim Report, vol. 1', report for the PWA, Bethlehem 2000 Committee and the WSSA (May 1998), pp. 10–11.
56 Schiff and Ya'ari, *Intifada*, p. 97. Schiff and Ya'ari are here reporting figures given by the one-time head of Israel's Civil Administration in the West Bank.

Notes on Chapter 4

1 I adopt this coupling of terms from Afif Safieh, *The Peace Process: From Breakthrough to Breakdown* (London: PLO, 1997).
2 US President Bill Clinton, Remarks at the Signing of the Israeli–PLO Declaration of Principles (Washington DC, 13 October 1993).
3 For a sample of such differing opinions, see David Makovsky, 'Oslo is not Dead. It Cannot Die', *Ha'aretz* (28 August 1998).
4 See Chomsky, *Fateful Triangle*, ch. 10; and Said, *Peace and its Discontents*. The title of Said's recent book – *The End of the Peace Process* – also of course invokes the vocabulary of breakdown, but does so purely idiomatically, and without any suggestion that the Oslo process represented a breakthrough in the Palestinian struggle for self-determination.

5 Said, *Peace and its Discontents*, p. 10.
6 A letter from Arafat to Rabin accompanying the 1994 Cairo Agreement stipulated that Arafat would 'use the title "Chairman (Ra'ees in Arabic) of the Palestinian Authority", or "Chairman of the PLO", and will not use the title "President of Palestine"' (4 May 1994).
7 Israel and the PLO, 'Interim Agreement on the West Bank and Gaza Strip' (Washington DC, 28 September 1995), Annex III, Appendix I, Article 29 (2); ibid., Article 28 (4).
8 Israel and the PLO, 'Agreement on Preparatory Transfer of Powers and Responsibilities' (Erez, 19 August 1994).
9 Israel and the PLO, 'Agreement on the Gaza Strip and the Jericho Area' (Cairo, 4 May 1994).
10 Israel and the PLO, 'Interim Agreement' – more commonly known as Oslo II.
11 Israel and the PLO, 'Protocol Concerning the Redeployment in Hebron' (Jerusalem, 17 January 1997); Israel and the PLO, 'Wye River Memorandum' (Washington DC, 23 October 1998); Israel and the PLO, 'Sharm-el-Sheikh Memorandum on Implementation of Outstanding Commitments of Agreements Signed and the Resumption of Permanent Status Negotiations' (Sharm-el-Sheikh, 4 September 1999). Figures from Geoffrey Aronson, 'Recapitulating the Agreements: The Israeli–PLO "Interim Agreements"', Centre for Policy Analysis on Palestine, Information Brief No 32 (27 April 2000).
12 Israel and the PLO, 'Interim Agreement', Article 18 (4.a).
13 Ibid., Article 17 (1.a).
14 Amnesty International, 'Israel and the Occupied Territories. Demolition and Dispossession: The Destruction of Palestinian Homes' (December 1999).
15 Israel and the PLO, 'Declaration of Principles', Article 8; 'Interim Agreement', Annex I, Article 4 (3.a,b).
16 The figure is from the Government of Israel, 'Palestinian Obligations as Per Note for the Record of the Hebron Protocol of January 15, 1997' (13 January 1998). Admittedly this is not the best source for impartial information, but is probably not too inaccurate. In a 1996-published article, the well-informed Graham Usher gave a figure of 30,000 for the PA police and security forces ('The politics of internal security: the PA's new intelligence services', *Journal of Palestine Studies*, vol. 25, no 2, 1996, p. 23), and their number would undoubtedly have grown after that. The claim regarding the proportion of Palestinian police was made by Raja Sourani in 1995, as reported in ibid., p. 34.
17 While estimates vary as to the number of PA police, the figure of 14 is given by the knowledgeable commentator Danny Rubinstein, 'Protection racket, PA-style', *Ha'aretz* (3 November 1999).
18 Israel and the PLO, 'Interim Agreement', Annex I, Article 4.2a.

19 Government of Israel, 'Palestinian Obligations as Per Note for the Record'.
20 Usher, 'The politics of internal security', pp. 27–8.
21 The CIA's role is set out in Israel and the PLO, 'Wye Memorandum', Article 2 (1.c).
22 See for instance Amnesty International, 'Palestinian Authority Silencing Dissent' (September 2000).
23 Israel and the PLO, 'Interim Agreement', Article 10.
24 On these economic dimensions, see Elmusa and Mahmud El-Jaafari, 'Power and trade: the Israeli–Palestinian economic protocol', *Journal of Palestine Studies*, vol. 24, no 2 (1995), pp. 14–32; Efrat, 'Interview with Adel Samara'; Emma Murphy, 'Stacking the deck: the economics of the Israeli–PLO accords', Middle East Report, vol. 25, no 3/4 (1995), pp. 35–8; Murphy, 'Israel and the Palestinians: The Economic Rewards of Peace', CMEIS Occasional Paper No 47 (March 1995); Peled, 'From Zionism to capitalism: the political economy of Israel's decolonization of the occupied territories', *Middle East Report*, no 194/5 (1995), pp. 13–17; Roy, *The Palestinian Economy and the Oslo Process: Decline and Fragmentation* (Abu Dhabi: Emirates Center for Strategic Studies and Research, 1998); Roy, 'De-development revisited: Palestinian economy and society since Oslo', *Journal of Palestine Studies*, vol. 28, no 3 (1999), pp. 64–82; Roy, 'Palestinian society and economy: the continued denial of possibility', *Journal of Palestine Studies*, vol. 30, no 4 (2001), pp. 5–20; and Adel Samara, 'Globalization, the Palestinian economy and the peace process', *Journal of Palestine Studies*, vol. 29, no 2 (2000), pp. 20–34.
25 Figure from Adel Samara in Efrat, 'Interview with Adel Samara'.
26 Asher Davidi, 'Israel's economic strategy for Palestinian independence', *Middle East Report*, no 184 (1993), p. 26.
27 Roy puts the figure, for mid-1998, at 89,000 ('De-development revisited', p. 71) while Samara, in an article published in 2000, puts the current figure at 100,000 ('Globalization, the Palestinian economy and the peace process', p. 24).
28 Israel and the PLO, 'Protocol on Economic Relations' (Paris, 29 April 1994), Article 5.
29 World Bank, 'Fifteen Months', p. vi.
30 This figure is extrapolated from Foundation for Middle East Peace, *Report on Israeli Settlement*, vol. 3, no 5 (September 1993), which put the West Bank settler population at 120,000; and ibid., vol. 10, no 5 (September 2000), which estimates the current settler population at 200,000.
31 LAW, 'Bypass Road Construction in the West Bank: The End of the Dream of Palestinian Sovereignty' (Jerusalem: LAW, 1996).
32 Baruch Kra, 'Settlement building up 81% in first quarter', *Ha'aretz* (22 August 2000), citing figures from the Israeli Central Bureau of Statistics.

33 Israel and the PLO, 'Declaration of Principles', Article 7 (4); Annex III (1).
34 Israel and the PLO, 'Agreement on the Gaza Strip and the Jericho Area', Annex II, Article 2 (B.31.a).
35 Israel and the PLO, 'Interim Agreement', Annex III, Appendix 1, Article 40 (1).
36 Ibid., Article 40 (11, 12).
37 Ibid., Article 40 (13, 14).
38 Ibid., Article 40 (12); Schedule 8.
39 Ibid., Article 40 (4); Schedule 8 (2.a, b).
40 Ibid., Article 40 (12); Schedule 8.
41 Ibid., Schedule 9 (1).
42 Ibid., Schedule 9 (2, 3).
43 Ibid., Schedule 9 (4).
44 Ibid., Article 40 (7).
45 Ibid., Article 40 (6).
46 Ibid., Schedule 10.
47 'Negotiators Achieve Breakthrough on Water Rights', Israel-Line (25 August 1995).
48 Elaine Fletcher, 'Israel, PLO Make Deal on West Bank Water', *The San Francisco Examiner* (21 September 1995); and citing her, Rouyer, *Turning Water into Politics*, p. 202.
49 Shapland, *Rivers of Discord*, p. 35; Rouyer, *Turning Water into Politics*, p. 207.
50 Dan Zaslavsky, Israeli Water Commissioner, 1990–92; cited in Rouyer, *Turning Water into Politics*, p. 206.
51 Sherman, *The Politics of Water in the Middle East*, pp. 108, 112–3.
52 Rouyer, *Turning Water into Politics*, ch. 7; and Kliot, 'A cooperative framework for sharing scarce water resources: Israel, Jordan, and the Palestinian Authority', in Amery and Wolf (eds), *Water in the Middle East*, p. 204.
53 Fadel Qawash, Deputy Head of the PWA, quoted in *Al Quds* (16 April 1998), p. 22; Qawash, PWA Press Conference (22 August 1998).
54 Rouyer, *Turning Water into Politics*, p. 213; Rouyer, 'Implementation of the water accords in the Oslo II Agreement', *Middle East Policy*, vol. 7 (1999), p. 113.
55 Interviews with Abdul Karim Asa'd, General Manager, Jerusalem Water Undertaking, Ramallah District, 20 July 1998; and Abdul Rahman Tamimi; also Tamimi, 'A Technical Framework for Final Status Negotiations over Water', *Palestine–Israel Journal of Politics, Economics and Culture*, vol. 3 (1996), pp. 70–2; Allegra Pacheco, 'Oslo II and Still No Water', Middle East International (3 November 1995), pp. 18–19; and Darci Vetter, 'The Impact of Article 40 of the Oslo II Agreement on Palestinian Water Provision' (Jerusalem: LAW, 1995).
56 Figures extrapolated from Israel and the PLO, 'Interim Agreement', Annex III, Appendix 1, Schedule 10.

57 Joseph Dellapenna, 'Developing a treaty regime for the Jordan Valley', in Eran Feitelson and Marwan Haddad (eds), *Joint Management of Shared Aquifers: The Fourth Workshop* (Jerusalem: Harry S. Truman Institute for the Advancement of Peace, and Palestine Consultancy Group, 1995), p. 207.
58 Israel and the PLO, 'Interim Agreement', Annex III, Appendix 1, Article 40 (18).
59 Interviews with Taher Nassereddin (12 April 1998), and Mohammed Jaas, West Bank Water Department (18 August 1999, and 1 June 2002).
60 Interview with Mustapha Nuseibi, Head, Hydrological Monitoring, West Bank Water Department (27 June 1998); also Water Resources Action Program, 'Hydrological Monitoring in Palestine: Status and Planning of the National Programme', Report for the Palestinian Water Authority (June 1996).
61 Interview with Mustapha Nuseibi.
62 Israel and the PLO, 'Interim Agreement', Annex III, Appendix 1, Schedule 9 (1).
63 Interviews with Mustapha Nuseibi, and Taher Nassereddin (15 August 1998).
64 Interview with Mustapha Nuseibi.
65 Interviews with Schmuel Kantor, Co-ordinator Special Duties, Water Commission, and Special Advisor, Mekorot (13 August 1998); Taher Nassereddin (15 August 1998); Mustapha Nuseibi; and Schmuel Zaslavsky, Chief Engineer, Mekorot, Ramle (24 August 1998); also Rouyer, *Turning Water into Politics*, p. 226.
66 Interview with Mustapha Nuseibi.
67 Ibid.
68 Interview with Basema Bashir, Manager, Hydrological Information System, PWA (4 April 1998).
69 For general discussion of this issue see Brynen, *A Very Political Economy*.
70 Interviews with Taher Nassereddin (12 April 1998 and 15 August 1998); and interview with leading Israeli water expert (to remain anonymous).
71 Interviews with Basema Bashir; and Schmuel Kantor.
72 For general discussion of this secrecy see especially Rouyer, *Turning Water into Politics*, pp. 15–17, 135–8.
73 Israel and the PLO, 'Interim Agreement', Annex III, Appendix 1, Article 40 (7), (7.b.vi).
74 Ibid., Article 40 (7).
75 Ibid., Article 40 (3.a).
76 Ibid., Schedule 10.
77 Interview with Yossi Guttman, Senior Hydrologist and Hydro-Geologist, Tahal (4 August 1998). Figure from Amjad Aliewi and Ayman Jarrar, 'Technical Assessment of the Potentiality of the Herodian Wellfield against Additional Well Development Programmes' Report for the PWA (April 2000), p. 6. This issue will be discussed in greater depth in the next chapter.

78 USAID West Bank and Gaza Mission, 'USAID-Funded Expansion of the Bethlehem-Hebron Water Supply System Complete', Press Release (5 December 1999).
79 Israel and the PLO, 'Interim Agreement', Annex III, Appendix 1, Schedule 8 (1.b). The same point is made by Elmusa, *Water Conflict*, p. 131; and Rouyer, *Turning Water into Politics*, p. 223. Bureaucratic details from interview with Mohammed Jaas (26 August 1999).
80 Interview with Taher Nassereddin (15 August 1998); Rouyer, *Turning Water into Politics*, pp. 225, 228.
81 Rouyer, *Turning Water into Politics*, pp. 228, 232; Schiff, 'Sharon suggests taking over water sources in West Bank', *Ha'aretz* (21 May 1997).
82 Interviews with Ayman Jarrar, PWA (17 August 1999), Omar Zayad, PWA (17 August 1999), and Mohammed Jaas (18 August 1999).
83 Interview with PA water official, to remain anonymous.
84 Israel and the PLO, 'Interim Agreement', Article 18 (4.a).
85 Interviews with Ayman Jarrar (17 August 1999), and Omar Zayad (17 August 1999); also Rouyer, *Turning Water into Politics*, pp. 225–6, 232; and Schiff, 'Sharon suggests taking over water sources in West Bank'.
86 On the early Netanyahu period I benefited from interviews with Karen Assaf, PWA (13 July 1998); Fadia Daibes, Deputy Programme Director, Norwegian Institutional Co-operation Programme, PWA (31 March 1998); Jad Isaac, ARIJ (28 March 1998); Nabil Sharif (11 February 1998); and Abdul Rahman Tamimi.
87 Formal accusations are contained in Government of Israel, 'Palestinian Obligations as Per Note for the Record', Annex V. For further coverage and discussion see L. Collins, 'Up from the Palestinian sewage spews politics: sewage from the Autonomous areas is threatening Israel's main aquifers', *Jerusalem Post* (6 December 1996), p. 7; Z. Hellman and P. Inbari, 'Water Pollution in the West Bank: Overcoming Political Stumbling Blocks to a Solution', *Institute for Peace Implementation* (29 April 1997), esp. p. 5; and Rouyer, *Turning Water into Politics*, p. 237. This issue was also discussed in interview with Karen Assaf, Schmuel Kantor, and Nabil Sharif. For an alternative take on the use of sewage as a political tool see Gideon Levy, 'The sewage of Ma'aleh Adumin', *Ha'aretz* (22 February 1998).
88 Interviews with Karen Assaf, PWA (13 July 1998 and 17 August 1999), Mohammed Jaas (26 August 1999 and 1 June 2002), and Taher Nassereddin (15 August 1998); also Rouyer, *Turning Water into Politics*, p. 247.
89 Interview (15 August 1998).
90 The PWA was formally established as a result of Presidential Decree No 90 for 1995 (issued 26 April 1995). Its form and functions were codified in PA, 'Law No 2 for 1996: Concerning the Establishment of the PWA' (issued by President Arafat, 18 January 1996).

91 Interview with Karen Assaf (1 June 2002).
92 Interviews with a number of Palestinian water experts, to remain anonymous; also Rouyer, *Turning Water into Politics*, p. 217.
93 Uri Savir, *The Process: 1,100 Days That Changed the Middle East* (New York: Random House, 1998), p. 215.
94 Rouyer, *Turning Water into Politics*, p. 207.
95 Interview with Mohammed Jaas (1 June 2002); also Noah Kinnarty, 'An Israeli view: if only they were quiet, the Palestinians have numerous opportunities', Interview, *Bitterlemons* (5 August 2002).
96 Israeli–Palestinian Joint Water Committee, 'Joint Declaration for Keeping the Water Infrastructure Out of the Cycle of Violence' (Erez, 31 January 2001).
97 Interview with Mohammed Jaas (1 June 2002).
98 Interview with Ayman Jarrar (1 June 2002).
99 *Fateful Triangle*, p. 538.
100 *Intimate Enemies*, p. 232.

Notes on Chapter 5

1 Israel and the PLO, 'Interim Agreement', Annex III, Appendix 1, Schedule 10.
2 Ibid., Article 40 (7.b.vi).
3 Ibid., Schedule 8 (1).
4 The latter figure is given in Schwarz, 'Water resources in Judaea and Samaria and the Gaza Strip', in J. D. Elazer (ed.), *Judaea, Samaria and Gaza* (Washington DC: American Enterprise Institute, 1982), p. 90; the former in Schwarz, 'Israel Water Sector Study: Past Achievements, Current Problems and Future Options', Report for the World Bank (1990), p. 11, and Ben Gurion University of the Negev and Tahal Consulting Engineers, 'Israel Water Study for the World Bank' (August 1994), p. 2.12.
5 Wolf, *Hydropolitics Along the Jordan River*, p. 10 adopts a figure of 125 mcmy as do Zarour and Isaac, 'The water crisis in the Occupied Territories', Paper presented at the VII World Conference on Water, Rabat, Morocco (12–16 May 1991). Rouyer, 'The water issue in the Israeli–Palestinian peace process', *Survival*, vol. 39, no 2 (1997), p. 60, and Elmusa, *Negotiating Water: Israel and the Palestinians* (Washington DC: Institute for Palestine Studies, 1996) suggest a yield of 120 mcmy. One Israeli–Palestinian collaborative venture put the aquifer's yield at 151 mcmy, 70 mcmy of which were labelled brackish (Karen Assaf et al., 'A Proposal for the Development of a Regional Water Master Plan', p. 30). Much earlier, two Israeli hydrologists had estimated the aquifer's yield to lie in the range of 123–174 mcmy: see Yohanan Boneh and Uri Baida, 'Water sources in Judea and Samaria and their exploitation', in

A. Shmueli et al. (eds), *Yehuda Veshomron* (Jerusalem: Kenaan, 1977–8), in Hebrew, p. 39.

6 Interviews with Yossi Guttman, Senior Hydrologist and Hydro-Geologist, Tahal (4 August 1998); and Ze'ev Golani, Mekorot (6 August 1998).
7 Interview with Yossi Guttman.
8 Interviews with Dvor Gillad, IHS (16 August 1998); Yossi Guttman; and David Scarpa, WSERU (30 March 1998).
9 Interviews with Dvor Gillad and Yossi Guttman.
10 Interview with Yossi Guttman.
11 Interviews with David Scarpa and Abdul Rahman Tamimi.
12 Interview with Yossi Guttman.
13 Aliewi and Jarrar, 'Technical Assessment of the Potentiality of the Herodian Wellfield against Additional Well Development Programmes', p. 6.
14 IHS, 'The Development, Exploitation and Condition of Groundwater Sources in Israel up to the Fall of 1997' (1998), p. 192.
15 Interviews with David Scarpa and Abdul Rahman Tamimi.
16 David Scarpa, personal communication (3 September 2000). For details see J. Kronfeld et al., 'Natural isotopes and water stratification in the Judea Group aquifer (Judean Desert)', *Israel Journal of Earth Sciences*, vol. 39 (1992), pp. 71–6; and Emanuel Mazor and Magda Molcho, 'Geochemical studies on the Feshcha Springs, Dead Sea basin', *Journal of Hydrology*, vol. 15 (1971), pp. 37–47.
17 Yossi Guttman, 'Hydrogeology of the Eastern Aquifer in the Judea Hills and the Jordan Valley', Report for the German–Israeli–Jordanian–Palestinian Joint Research Program for the Sustainable Utilization of Aquifer Systems (18 January 1998).
18 Interview with David Scarpa (25 August 1998).
19 Interviews with Yossi Guttman and Abdul Rahman Tamimi.
20 Interview with Yossi Guttman.
21 Wolf, *Hydropolitics Along the Jordan River*, p. 9.
22 Oscar Meinzer, 'Outline of Groundwater Hydrology with Definitions', US Geological Survey, *Water Supply Paper 494* (1923), p. 55.
23 David Todd, *Groundwater Hydrology* (New York: John Wiley and Sons, 1959), p. 200.
24 Ibid., p. 363; Charles Fetter, *Applied Hydrogeology*, 3rd edn. (New Jersey: Prentice Hall, 1994), p. 519.
25 Interview with Joshua Schwarz.
26 Figure from interview with Dvor Gillad.
27 See for instance CDM/Morganti, 'Task 26: Environmental Assessment for the Hebron-Bethlehem Water Supply System, Final Environmental Assessment', Report for the PWA (18 May 1997).
28 CDM/Morganti, 'Task 20', p. 4.
29 Interview with Hisham Zarour, Head, Water Resources and Planning Department, PWA (10 August 1998). The reports in question are Rofe

and Raffety Consulting Engineers, 'Jerusalem and District Water Supply, Geological and Hydrological Report', Report for the Central Water Authority, Hashemite Kingdom of Jordan (1963); 'Nablus District Water Resources Survey, Geological and Hydrological Report', Report for the Central Water Authority, Hashemite Kingdom of Jordan (1965); 'West Bank Hydrology, 1963–65', Report for the Central Water Authority, Hashemite Kingdom of Jordan (1965).

30 Lonergan and Brooks, *Watershed*, p. 147. Sharif Elmusa's *Water Conflict* provides a notable exception to such information deficit thinking.

31 Interview with Yossi Guttman.

32 Aliewi and Anan Jayyousi, 'The Palestinian Water Resources in the Final Status Negotiations: Technical Framework and Professional Perception', Report for the PWA (4 May 2000), p. 14.

33 For further speculation on these matters see Elmusa, *Water Conflict*, pp. 38–42.

34 See for instance CDM/Morganti, 'Comprehensive Planning Framework for Palestinian Water Resources Development, Final Report', vol. 2, Report for the PWA (6 July 1997); GTZ, 'Middle East Regional Study on Water Supply and Demand Development, Phase 1 Report'; MOPIC, 'Regional Plan for West Bank Governorates: Water and Wastewater' (July 1998).

35 I allude here to Latour, *Science in Action: How to Follow Scientists and Engineers Through Society* (Cambridge MA: Harvard University Press, 1987), especially pp. 2–3.

36 CDM/Morganti, 'Task 18: Study of the Sustainable Yield of the Eastern Aquifer Basin. Final Report' (1998).

37 Interview with Hisham Zarour.

38 Aliewi and Jarrar, 'Technical Assessment of the Potentiality of the Herodian Wellfield against Additional Well Development Programmes', p. 3.

39 PHG, 'Water for Palestine: A Critical Assessment of the European Investment Bank's Lending Strategy in the Rehabilitation of Water Resources in the Southern West Bank', Report for the Reform the World Bank Campaign (2000), pp. 8, 6.

Notes on Chapter 6

1 Jane Corbin emphasises this personal dimension in her *Gaza First: The Secret Norway Channel to Peace between Israel and the PLO* (London: Bloomsbury, 1994), as do Uri Savir in *The Process*, and from the Palestinian side of the negotiations, Mahmud Abbas, *Through Secret Channels* (Reading: Garnet, 1995). Corbin's emphasis is explicitly critiqued by John King, *Handshake in Washington: The Beginning of Middle East Peace?* (London: Ithaca, 1994).

2 Presentation before the Peace and Security Council, May 23 1998; quoted in Ze'ev Schiff, *Security for Peace: Israel's Minimal Security Requirements in Negotiations with the Palestinians* (Washington DC: Washington Institute for Near East Policy, 1989), pp. 15–16.
3 Chomsky, *Fateful Triangle*, pp. 47–50.
4 For the text, see Walter Laqueur and Barry Rubin (eds), *The Israel–Arab Reader: A Documentary History of the Middle East Conflict* (New York: Penguin, 1984), pp. 609–15.
5 Chomsky, *World Orders, Old and New* (London: Pluto, 1994), pp. 231–8; William Quandt, *American Diplomacy and the Arab–Israeli Conflict Since 1967* (Berkeley: University of California Press, 1993), pp. 388–92; Charles Smith, *Palestine and the Arab–Israeli Conflict*, 3rd edn. (New York: St Martin's Press, 1996), p. 304.
6 Chomsky, *Fateful Triangle*, p. 527.
7 Shafir and Peled, *Being Israeli.*
8 On the economic transformations see for instance Kleiman, 'The waning of Israeli etatism', *Israel Studies*, vol. 2, no 2 (1988), pp. 146–71; and Shalev, 'Have globalisation and liberalisation "normalised" Israel's political economy?', *Israel Affairs*, vol. 5, nos 2–3 (1998), pp. 121–55.
9 Dov Lautman in *Davar*, 17 February 1993; quoted in Davidi, 'Israel's economic strategy for Palestinian independence', p. 24.
10 Shafir and Peled, *The New Israel: Peacemaking and Liberalization* (Boulder: Westview, 2000); Shafir and Peled, *Being Israeli*, chs 8 and 9.
11 Usher, *Dispatches from Palestine*, p. 43.
12 Schiff, *Security for Peace*, p. 53.
13 Tigva Honig-Parnass, 'A new stage: military intifada, Israeli panic, ruthless repression', *News From Within* (2 April 1993) pp. 1–4.
14 *Ha'aretz* (7 September 1993); quoted in Usher, *Dispatches from Palestine*, p. 74.
15 D. Hoffman, 'Shamir plan was to stall autonomy: Rabin says he'll cut subsidies to settlers', *Washington Post* (27 June 1992); quoted in Smith, *Palestine and the Arab–Israeli Conflict*, p. 314.
16 For a classic exposition of this argument see Chomsky, *Fateful Triangle*, pp. 44–54.
17 UN Security Council Resolution 242 (22 November 1967) is reproduced in Laqueur and Rubin, *The Israel–Arab Reader*, pp. 365–6. On the developing PLO policy see for instance Alain Gresh, *The PLO: The Struggle Within* (London: Zed, 1988).
18 On this distinction see especially Chomsky, *Fateful Triangle*, ch. 3.
19 Israel and the PLO, 'Declaration of Principles', Article 1. For a clear summary of the Israeli and US positions on 242 see Smith, *Palestine and the Arab–Israeli Conflict*, p. 211–3.
20 'Rabin: let's talk about success, not failure', interview with David Makovsky, *Jerusalem Post*, International edn (16 October 1993); quoted in Smith, *Palestine and the Arab–Israeli Conflict*, p. 324.

21 Smith, *Palestine and the Arab–Israeli Conflict*, pp. 257–8.
22 Savir, *The Process*, pp. 67, 204.
23 *Davar* (11 November 1982); quoted in Chomsky, *Fateful Triangle*, p. 112.
24 Quoted in Chomsky, *Fateful Triangle*, p. 542.
25 For an analysis, see Walid Khalidi, 'The Gulf crisis: origins and consequences', *Journal of Palestine Studies*, vol. 20, no 2 (1991), pp. 5–28.
26 Smith, *Palestine and the Arab–Israeli Conflict*, p. 312.
27 Samih Farsoun and Jean Landis, 'The sociology of an uprising: the roots of the *intifada*', in Nassar and Heacock, *Intifada*, pp. 15–35, and Helga Baumgarten, '"Discontented people" and "outside agitators": the PLO in the Palestinian uprising', in ibid., pp. 207–26; also Hillel Frisch, *Countdown to Statehood: Palestinian State Formation in the West Bank and Gaza* (Albany: State University of New York Press, 1998), ch. 5.
28 This status was granted the PLO by the Arab League in 1974; for the text see Laqueur and Rubin, *The Israel–Arab Reader*, p. 518. On Hamas see for instance Ziad Abu-Amr, 'Hamas: a historical and political background', *Journal of Palestine Studies*, vol. 22, no 4 (1993), pp. 5–19; Hisham Ahmad, *Hamas: From Religious Salvation to Political Transformation: The Rise of Hamas in Palestinian Society* (Jerusalem: PASSIA, 1994); and Usher, *Dispatches from Palestine*, ch. 2.
29 See the comments by two of the leading members of the Palestinian delegation to Madrid: Haider Abd al-Shafi, 'The Oslo Agreement: an interview with Haider Abd al-Shafi', *Journal of Palestine Studies*, vol. 22, no 1 (1993), pp. 14–19; and Hanan Ashrawi, *This Side of Peace* (New York: Simon and Schuster, 1995), esp. pp. 260–1.
30 Connie Bruck, 'The wounds of peace', *New Yorker* (14 October 1996); quoted in Nick Guyatt, *The Absence of Peace: Understanding the Israeli–Palestinian Conflict* (London: Zed, 1998), p. 43.
31 Allon Groth, *The PLO's Road to Peace: Processes of Decision-Making* (London: Royal United Services Institute for Defence Studies, 1995).
32 Chomsky, 'The Israel–Arafat agreement', *Z Magazine* (October 1993); reproduced in *Fateful Triangle*, pp. 533–40.
33 Ahmad Khalidi makes this point well in 'Security in the final Middle East settlement: some components of Palestinian national security', *International Affairs*, vol. 71, no 1 (1995), pp. 1–18.
34 Savir, *The Process*, p. 14.
35 *Ha'aretz* (14 November 1994); quoted in Emma Murphy, 'Stacking the deck', p. 36.
36 *Ma'ariv* (25 October 1995); quoted in Jan de Jong, 'The geography of politics: Israel's settlement drive after Oslo', in George Giacaman and Dan Jorund Lonning (eds), *After Oslo: New Realities, Old Problems* (London: Pluto, 1998), p. 77.
37 Isaac, 'Water and Palestinian–Israeli peace negotiations', *Palestine Center: News and Analysis* (19 August 1999). See also for example Zarour and

Isaac, 'Nature's apportionment and the open market'.

38 Cecilia Albin, 'When the weak confront the strong: justice, fairness and power in the Israeli–PLO Interim talks', *International Negotiation*, vol. 4, no 2 (1999), p. 339.

39 Rouyer, *Turning Water into Politics*, p. 193.

40 Interview with Noah Kinnarty.

41 Quoted in L. Collins, 'Water rights negotiations move along a very slippery road', *Jerusalem Post* (29 July 1995).

42 S. Libiszewski, 'Water Disputes in the Jordan Basin Region and their Role in the Resolution of the Arab–Israeli Conflict', Environment and Conflicts Project, Occasional Paper No 13 (Zurich: Center for Security Studies and Conflict Research, 1995), pp. 82–5; Peters, *Pathways to Peace*, pp. 16–22; Rouyer, *Turning Water into Politics*, p. 185.

43 Albin, 'When the weak confront the strong', p. 339.

44 Savir, *The Process*, p. 214.

45 Rouyer, *Turning Water into Politics*, p. 195.

46 Peretz Kidron, 'A bomb hardens the mood', *Middle East International* (25 August 1995); Schiff, 'Water dispute deferred', *Ha'aretz* (8 September 1995); Feitelson, 'The ebb and flow of the Arab–Israeli water conflict', p. 360.

47 Their various ideas and proposals have been developed through a series of workshops and publications: Feitelson and Haddad (eds), *Joint Management of Shared Aquifers: The First Workshop* (1994); *Joint Management of Shared Aquifers: Final Report* (1995); *Joint Management of Shared Aquifers: The Fourth Workshop* (1998); Haddad and Feitelson (eds), *Joint Management of Shared Aquifers: The Second Workshop* (1995); and *Joint Management of Shared Aquifers: The Third Workshop* (1997); all Jerusalem: Palestine Consultancy Group and Harry S. Truman Institute for the Advancement of Peace.

48 Interviews with Noah Kinnarty and Moshe Yizraeli.

49 Interview with Noah Kinnarty; also Savir, *The Process*, p. 213.

50 Interviews with Schmuel Kantor, Co-ordinator Special Duties, Water Commission, Senior Advisor, Mekorot (13 August 1998); and Moshe Yizraeli.

51 Interview with Noah Kinnarty.

52 Interviews with Abdul Rahman Tamimi; Abdelkarim Asa'd; Marwan Haddad; and Anan al-Jayyousi, Director WESC, An-Najah University (14 April 1998).

53 Rouyer, *Turning Water into Politics*, p. 186; Savir, *The Process*, p. 214.

54 Interviews with Schmuel Kantor, Noah Kinnarty, and several Palestinian water experts.

55 Interview with Marwan Haddad.

56 Albin, 'When the weak confront the strong', p. 341; 'Negotiators achieve breakthrough on water rights', *Israel-Line* (25 August 1995); Ian Scobbie,

'H_2O after Oslo II: legal aspects of water in the Occupied Territories', in *Palestine Yearbook of International Law*, 1995/5, vol. 8 (Nicosia: Al-Shaybani Society of International Law, 1996), p. 104; Usher, 'Squeezing out the last drop', *Middle East International* (8 September 1995), p. 6.

57 Conversations with various Israeli and Palestinian water experts, to remain anonymous.

58 Interview with Abdul Rahman Tamimi.

59 Said, *Peace and its Discontents*, pp. 13–14.

60 Finkelstein, 'The real meaning of the Wye River Memorandum' (28 July 1998).

61 On the PLO in this regard see especially Frisch, *Countdown to Statehood*; Sayigh, 'Armed struggle and state formation', *Journal of Palestine Studies*, vol. 26, no 4 (1997), pp. 17–32.

62 Foucault, *Discipline and Punish*, p. 139; Said, *Peace and its Discontents*, p. 25.

63 Interviews with Ayman Jarrar (17 August 1999), Omar Zayad (17 August 1999) and Mohammed Jaas (18 August 1999).

64 Albin, 'When the weak confront the strong', pp. 336, 346, 347; and see also Albin, *Justice and Fairness in International Negotiation* (Cambridge: Cambridge University Press, 2001), ch. 5.

Notes on Chapter 7

1 Migdal, *Through the Lens of Israel*, especially ch. 2

2 For accounts of this system, see esp. Brynen, 'The dynamics of Palestinian elite formation', *Journal of Palestine Studies*, vol. 24, no 3 (1995), pp. 31–43; Brynen, 'The neopatrimonial dimension of Palestinian politics', *Journal of Palestine Studies*, vol. 25, no 1 (1995), pp. 23–36; Frisch, *Countdown to Statehood*, ch. 7; George Giacaman, 'In the throes of Oslo: Palestinian society, civil society and the future', in Giacaman and Lonning, *After Oslo*, pp. 1–15; Jamil Hilal, 'The effect of the Oslo Agreement on the Palestinian political system', in ibid., pp. 121–45; Christopher Parker, *Resignation or Revolt: Socio-Political Development and the Challenges of Peace in Palestine* (London: I.B.Taurus, 1999); Glen Robinson, 'The growing authoritarianism of Arafat's regime', *Survival*, vol. 39, no 2 (1997), pp. 42–56; Usher, *Palestine in Crisis*; and Usher, *Dispatches from Palestine*.

3 See for instance Usher, 'The rise of political Islam in the occupied territories', *Middle East International* (25 June 1993), pp. 19–20.

4 Roy, 'Gaza: new dynamics of civil disintegration', *Journal of Palestine Studies*, vol. 22, no 4 (1993), pp. 22–37.

5 On these inside-outside dynamics see Frisch, *Countdown to Statehood*, and Sayigh, 'Armed struggle and state formation'.

6 Usher, 'Palestinian trade unions and the struggle for independence', *Middle East Report*, no 194/5 (1995), pp. 20–23.

7 Frisch, *Countdown to Statehood*, ch. 5.
8 See especially Joost Hilterman, *Behind the Intifada: Labour and Women's Movements in the Occupied Territories* (Princeton: Princeton University Press, 1991), and Glen Robinson, *Building a Palestinian State: The Incomplete Revolution* (Bloomington: Indiana University Press, 1997).
9 Hilterman, *Behind the Intifada*, esp. pp. 65–6.
10 Roy, *The Gaza Strip: The Political Economy of De-Development.*
11 See for example on business taxes, Salim Tamari, 'The revolt of the petite bourgeoisie: urban merchants and the Palestinian uprising', in Nassar and Heacock, *Intifada*, pp. 159–73.
12 Frisch, *Countdown to Statehood*, p. 133.
13 Usher, *Palestine in Crisis*, pp. 66–7.
14 Frisch, *Countdown to Statehood*, ch. 7; and Usher, *Palestine in Crisis*, ch. 6.
15 Usher, *Palestine in Crisis*, p. 61.
16 See for instance Ziad Abu-Amr, 'The Palestinian Legislative Council: a critical assessment', *Journal of Palestine Studies*, vol. 26, no 4 (1997), pp. 90–97; Ilan Halevi, 'Self-government, democracy and mismanagement under the Palestinian Authority', *Journal of Palestine Studies*, vol. 27, no 3 (1998), pp. 35–48; and Usher, 'Arafat's new cabinet – "back me or sack me"', *Middle East International* (21 August 1998), pp. 3–4.
17 Brynen, 'The neopatrimonial dimension of Palestinian politics'. This Weberian concept was developed by Eisenstadt, *Traditional Patrimonialism and Modern Neo-Patrimonialism* (London: Sage, 1973).
18 R. Bergman, 'Israel deposited NIS 1.5b in Arafat's personal account', *Ha'aretz* (8 October 1999); Bergman, 'How much PA corruption is too much?' *Ha'aretz* (19 October 1999); Amira Hass, 'Chairman Arafat straightens out his financial accounts', *Ha'aretz* (13 January 2000).
19 For analyses of the international assistance programme see Brynen, 'International aid to the West Bank and Gaza'; Brynen, 'Buying peace? A critical assessment of international aid to the West Bank and Gaza', *Journal of Palestine Studies*, vol. 25, no 3 (1996), pp. 79–92; Brynen, *A Very Political Economy*; Center for Palestine Research and Studies, 'Prevailing Perceptions of Aid Management', Research Report No 9 (December 1997); JMCC, 'Foreign Aid and Development in Palestine, Phase I Report' (1997); JMCC, 'Mortgaging Self-Reliance: Foreign Aid and Development in Palestine, Phase II Report' (1997); JMCC, 'Foreign Aid and Development in Palestine, Phase III Report' (1999).
20 Brynen, 'Buying Peace?'.
21 Samara, 'Globalization, the Palestinian economy and the "peace process"', p. 24.
22 Brynen, *A Very Political Economy*, pp. 187–91.
23 Ron Efrat, 'Porcupine tangos: the PA and the NGOs (accompanied by a CIA orchestra)', *Challenge*, no 57; Nina Sovich, 'Arafat's watch on Palestinian NGOs', *Palestine Report* (5 February 1999); Dennis Sullivan,

'NGOs in Palestine: agents of development and foundation of civil society', *Journal of Palestine Studies*, vol. 25, no 3 (1996), pp. 93–100.

24 Quoted in Brynen, *A Very Political Economy*, p. 191.

25 Interview with Karen Assaf (1 June 2002); see also Rouyer, *Turning Water into Politics*, p. 247.

26 Interviews with Fadia Daibes; and Jane Meyer, Programme Management Officer, UNDP (24 July 1998); also Rouyer, *Turning Water into Politics*, p. 216.

27 Interviews with several PWA and other Palestinian water experts, to remain anonymous.

28 Interview with PWA official, to remain anonymous.

29 Interviews with several Palestinian water experts, to remain anonymous.

30 Elmusa, *Water Conflict*, p. 273.

31 PWA, 'Internal Regulations'; and interview and conversations with Fadia Daibes.

32 PA, Law no 2 for 1996.

33 Interview with Fadia Daibes.

34 PA, Law no 2 for 1996; PWA, 'Internal Regulations' (1996); interview with Fadia Daibes; and Rouyer, *Turning Water into Politics*, pp. 217–9.

35 Interview with Fadia Daibes.

36 Letter from Nabil Sharif, PWA Chairman to Odin Knudsen, World Bank Representative for the West Bank and Gaza, 'Re: Gaza Water and Wastewater Services Project' (13 January 1996).

37 Interview with PWA official, to remain anonymous.

38 Interviews with Anan Al-Jayyousi, Abdul Rahman Tamimi, and many other Palestinian water experts, including PWA officials. Rouyer, *Turning Water into Politics*, p. 214, also reports this as a view held by PA officials.

39 Amongst the facility and district-level plans for the southern West Bank are CDM/Morganti, 'Task 25: Water Supply Facility Plan for the Hebron-Bethlehem Service Area'; CDM/Morganti, 'Task 33: Integrated Water Resources Management Plan for the Hebron-Bethlehem Area'; and SOGREAH Ingenierie, 'Master Plan for Water Distribution in the Bethlehem Area'. Parallel plans have been developed, above all by CDM/Morganti, for the Jenin, Nablus and Ramallah areas of the West Bank. Among the regional plans for the West Bank are MOPIC, 'Regional Plan for West Bank Governorates: Water and Waste Water' (July 1998), discussed further below. Among the master plans and strategic plans are WRAP, 'Palestinian Water Resources: A Rapid Interdisciplinary Sector Review and Issues Paper' (October 1994), produced prior to the establishment of the PWA; CDM/Morganti, 'Task 4: Comprehensive Planning Framework for Palestinian Water Resource Development'; World Bank, 'Strategic Water Resources Plan and Priority Investment Identification', Report for the PWA (1999); among the multilateral planning projects have been GTZ, 'Middle East Regional Study on Water Supply and

Demand Development'; see for instance the 'Concluding Report' (February 1998).

40 MOPIC, 'Regional Plan: Water and Waste Water'. The plan was produced by Dorsch Consult of Munich, in association with Universal Engineering Consulting of Ramallah. I was fortunate enough to attend the final presentation of the plan by German consultants Dorsch Consult to MOPIC and PWA officials on 28 July 1998, two days before it was to be officially submitted to MOPIC. The following account is based on this presentation and the arguments that quickly ensued, as well as on interviews and conversations with Karen Assaf; Natasha Carmi, MOPIC (31 March 1998); Ibrahim Dajani, MOPIC (various 1998); Frank Haugan, MOPIC consultant (various 1998); Olal Hauge, MOPIC consultant (various 1998); Ali Al-Labadi, Universal Engineering (31 March 1998); Hiba Tahboub; Bader Zahrer, MOPIC (31 March 1998); and Hisham Zarour.

41 Interview with Jane Meyer, UNDP (24 July 1998).

42 Such was the explicit purpose of the World Bank-funded 'Strategic Water Resources Plan', developed during 1998–9, for instance (see World Bank, 'Strategic Water Resources Plan: Outline Terms of Reference, Draft', 16 March 1996). On this I benefited from interviews with Anan Al-Jayyousi, Jane Meyer, and Hiba Tahboub, and several Palestinian water experts and international workers (August 1999); I was also present at the launch of the project in Ramallah (27 July 1998).

43 Interviews with several PWA officials and international contractors, to remain anonymous; also Trottier, *Hydropolitics in the West Bank and Gaza Strip*, p. 96.

44 Save the Children Fund West Bank and Gaza, 'Project Submission to DfID for Water and Sanitation Program in the Dura Area, Southern West Bank' (August 1998).

45 Interview with Taher Nassereddin.

46 Interview with Mohammed Jass (1 June 2002).

47 Interviews with Taher Nassereddin and Fadia Daibes.

48 Interview with Taher Nassereddin (12 April 1998).

49 The following account is based on informal conversations with residents of Duwarra, July and August 1998; and interview with Taher Nassereddin (15 August 1998).

50 In interview, Taher Nassereddin denied that the Water Department co-operated with Kiryat Arba; the residents of Duwarra were, however, adamant that this was the case.

51 Interview with Taher Nassereddin (15 August 1998).

52 Conversation with Palestinian water engineer, to remain anonymous.

53 Villager leaders were adamant that no notification was received; Taher Nassereddin claims the reverse.

54 Interview with Mohammed Laham.

55 Trottier, *Hydropolitics in the West Bank and Gaza*, pp. 74–7.
56 Interviews with Mahmud Abu-Deyah (19 August 1999), Andre Naudy and Walid Shalan, and informal conversations with many other Palestinians in Bethlehem, Beit Jala and Dheisheh during summers 1998 and 1999.
57 Interviews with Mahmud Abu-Deyah and Andre Naudy.
58 Interview with Mahmud Abu-Deyah; also 'PA cracks down on debtors', *Middle East Newsline* (26 September 2000).
59 Hass, 'A report on Palestinian water crisis', *Al Quds* (25 June 1999); reproduced in CDM/Morganti, *H_2O* (June 1999), p. 6.
60 Interviews with Ayman Jarrar (1 June 2002) and Ihab Bargouthi (2 June 2002).
61 CDM/Morganti, *Task 25, Final Report*, p. 5.
62 Interview with Ihab Bargouthi (2 June 2002).
63 USAID West Bank and Gaza Mission, 'Water Produced From the First Palestinian Owned and Operated Well in Bethlehem-Hebron Water Supply System', Press Release (26 July 1999); CDM/Morganti, *H_2O* (30 September 1998), pp. 1–2; and interview with Mustapha Nuseibi (27 June 1998).
64 USAID website at http://www.usaid-wbg.org/water.html.
65 Israel and the PLO, 'Interim Agreement', Annex III, Appendix 1, Schedule 8 (2.a).
66 USAID West Bank and Gaza Mission, 'Water Produced from the First Palestinian Owned and Operated Well in Bethlehem-Hebron Water Supply System'; and interview with Ihab Bargouthi, CDM/Morganti and PWA (18 August 1999).
67 CDM/Morganti, *H_2O* (30 September 1996), p. 2; USAID West Bank and Gaza Mission, 'USAID Contracts to Help Improve West Bank Water Systems', Press Release (17 July 1996); 'USAID to Build New Water Supply Systems for the West Bank', Press Release (14 January 1998); and interview with Qasem Awad, Resident Engineer, CDM/Morganti (26 August 1998).
68 CDM/Morganti, *H_2O* (30 September 1997), p. 5; CDM/Morganti, *H_2O* (31 March 1998), p. 1; CDM/Morganti, *H_2O* (30 September 1998), p. 2; and meeting with Qasem Awad.
69 CDM/Morganti, *H_2O* (31 March 1998), p. 1.
70 Interview with Ihab Bargouthi.
71 Interviews with Qasem Awad, Ihab Bargouthi and Ayman Jarrar.
72 Interview with Ihab Bargouthi.
73 Interview with international consultant, to remain anonymous.
74 CDM/Morganti, *H_2O* (30 June 1999), p 1; and interview with Qasem Awad.
75 Interview with international consultant, to remain anonymous.
76 Interview with Ihab Bargouthi.
77 Interviews with Ayman Jarrar (1 June 2002) and Ihab Bargouthi (2 June 2002).

78 CDM/Morganti, *H_2O* (30 September 1998), p. 2. This account draws upon interviews with Qasem Awad, Ihab Bargouthi (12 August 1998), and other Palestinian water experts, to remain anonymous.
79 This account draws primarily upon interviews with Karen Assaf, Noah Kinnarty, Andre Naudy, Mustapha Nuseibi, David Scarpa and Nabil Sharif; see also Trottier, *Hydropolitics in the West Bank and Gaza Strip*, pp. 94–7; and Trottier, 'Water and the challenge of Palestinian institution building', pp. 46–7.
80 CDM/Morganti, *H_2O* (31 March 1997), p. 3.
81 Israel and the PLO, 'Interim Agreement', Annex III, Appendix 1, Article 40 (7.a.i). Water supply details from interviews with Mohammed Jaas and Mustapha Nuseibi.
82 Interviews with Ihab Bargouthi (2 June 2002) and Mahmud Abu Deyah (4 June 2002).
83 Interview with Ihab Bargouthi (2 June 2002).
84 Interview with Ihab Bargouthi (18 August 1999, 2 June 2002).
85 Interview with Marie-Louise Weighill, Save the Children UK (4 June 2002).
86 Interview with Ihab Bargouthi (2 June 2002).

Notes on Chapter 8

1 Note for instance the recurring titles: on the Palestinian territories, A. Bellisari, 'Public health and the water crisis in the occupied Palestinian territories', *Journal of Palestine Studies*, vol. 23, no 2 (1994), pp. 52–63; Isaac, 'A Palestinian perspective on the water crisis', *Israel–Palestine Journal of Politics, Economics and Culture*, vol. 5, no 1 (1998), pp. 54–8; Isaac and Selby, 'The Palestinian water crisis'; Zarour and Isaac, 'The water crisis in the Occupied Territories'; on the Middle East as a whole, George Gruen, *The Water Crisis: The Next Middle East Crisis?* (Los Angeles: Wiesenthal Center, 1991); R. Sexton, 'The Middle East water crisis: is it the making of a new Middle East regional order?' *Capitalism, Nature, Socialism*, vol. 3, no 4 (1992), pp. 65–77; P. Vesilind, 'The Middle East's water: critical resource', *National Geographic*, vol. 183, no 5 (1993), pp. 38–71; World Bank, 'From Scarcity to Security: Averting a Water Crisis in the Middle East and North Africa'; and on a global level, R. Clarke, *Water: The International Crisis* (London: Earthscan, 1993); Peter Gleick, *Water in Crisis: A Guide to the World's Fresh Water Resources* (New York: Oxford University Press, 1993); and Fred Pearce, *The Damned: Rivers, Dams and the Coming World Water Crisis* (London: Bodley Head, 1992).
2 Allan, *The Middle East Water Question*, p. 173.
3 Ibid., pp. 174, 177; Allan, 'Striking the "right price" for water? Achieving harmony between basic human need, available resources and commercial viability', in Allan and Mallat, *Water in the Middle East*, p. 325.

4 Allan, *The Middle East Water Question*, p. 30.
5 Ibid., p. 177.
6 Allan, 'Striking the "right price" for water?' p. 344.
7 Chibli Mallat, 'The quest for water use principles: reflections on Shari'a and custom in the Middle East', in Allan and Mallat, *Water in the Middle East*, pp. 127–37.
8 Ibid., p. 174.
9 For some excellent accounts of these see Trottier, *Hydropolitics in the West Bank and Gaza Strip*, especially pp. 105–17; and Trottier, 'Water and the challenge of Palestinian institution building'.
10 This account is based on ethnographic research, informal conversations and interviews conducted on a large number of occasions during spring and summer 1998, and August 1999. Besides the ethnographic work, I should mention in particular interviews with Ziad Abbas, Dheisheh Community Centre (30 March 1998); Mahmud Abu-Deyah, Senior Engineer, WSSA (16 July and 25 August 1998, and 19 August 1999); Muna Hamzeh-Muhaisen, journalist and camp resident (17 August 1998); Mohammed Laham, Director, Camp Services Committee, Dheisheh (various 1998 and 1999); Tigritte Laham, Palestinian Ministry of the Environment (various 1998 and 1999); and Husayn Shaheen, Camp Services Director, Dheisheh (16 August 1999). Many thanks also to Yasser-Abid Alqfar, for his insights and help in and around Dheisheh. The water situation in Dheisheh was often reported on during 1998 and 1999; see for example Hamzeh-Muhaisen, 'Who's to blame for the severe water shortages?' Hass, 'Dire water shortages in West Bank', *Ha'aretz* (27 July 1998); Hass, 'Cut and dried'; L. King, 'Mideast neighbors quarrel over water', *Associated Press* (7 August 1999); and N. Shiyoukhi, 'Palestinians protest water shortage', *Associated Press* (17 July 1999).
11 Figure from http://www.geocities.com/CapitolHill/9836/dheisheh/camp.html.
12 Conversation with camp resident (3 July 1998).
13 This account is based largely on impressions gleaned from living in Beit Jala during the summers of 1998 and 1999.
14 My understanding of the situation in Quasiba is based on ethnographic research and conversations conducted on a number of occasions during the summer of 1998, as well as on the personal experience of living in the village during the autumn of 1994.
15 Interview with Abdul Rahman Tamimi. The same point is made by Rita Giacaman in her *Life and Health in Three Palestinian Villages* (London: Ithaca, 1988), p. 77.
16 Trottier, *Hydropolitics in the West Bank and Gaza Strip*, pp. 116–7.
17 Allan, *The Middle East Water Question*, p. 4.
18 My emphasis here is strongly informed by Michel de Certeau, *The Practice of Everyday Life* (Berkeley: University of California Press, 1984).

19 8 August 1998.
20 17 August 1998.
21 On the latter see for instance Bellisari, 'Public health and the water crisis in the occupied Palestinian territories'.

Notes on Conclusion

1 I owe this phrase to Benvenisti, *Intimate Enemies*, p. 218.
2 Finkelstein, 'Whither the peace process', *New Left Review*, no 218 (July/August 1996), p. 147.
3 Ibid., p. 145.
4 Chomsky, *Fateful Triangle*, p. 559.
5 On the water negotiations I benefited from an interview with Shaddad Attili, Negotiation Affairs Department, PA (2 June 2002).
6 Secretary of State Madeline Albright, quoted in Donald Neff, 'The US piles on the pressure', *Middle East International*, no 631 (18 August 2000), p. 4.
7 Foundation for Middle East Peace, *Report on Israeli Settlement in the Occupied Territories*, vol. 10, no 5 (September 2000).
8 Robert Malley and Hussein Agha, 'Camp David: the tragedy of errors', *New York Review of Books* (9 August 2001).
9 Usher, 'Running out of options?' *Middle East International* (1 September 2000), p. 4.
10 Agha and Malley, 'Camp David and after: a reply to Ehud Barak', *New York Review of Books* (13 June 2002); 'Moratinos Document' reprinted in *Ha'aretz* (14 February 2002). As Akiva Eldar notes, this document was 'approved by the Taba negotiators as an accurate description of the [January 2001] discussions', *Ha'aretz* (14 February 2002). For discussion of these post-Camp David negotiations, see especially Deborah Sontag, 'Quest for Mideast peace: how and why it failed', *New York Times* (26 July 2001). This article, and the earlier piece by Malley and Agha, became the subject of a rebuttal by Ehud Barak and leading Israeli historian Benny Morris, and of a subsequent debate with Malley and Agha; see Morris, 'Camp David and after: an interview with Ehud Barak', *New York Review of Books* (13 June 2002); Agha and Malley, 'Camp David and after'; Morris, Barak, Agha and Malley, 'Camp David and after – continued', *New York Review of Books* (27 June 2002). As any quick perusal of this exchange should make clear, Malley and Agha's and Sontag's versions of events are much more credible than those of Barak and Morris.
11 Officially the 'Report of the Sharm El-Sheikh Fact-Finding Committee', chaired by former US Senator George Mitchell (30 April 2001).
12 Khalil Shikaki, 'Old Guard, Young Guard: The Palestinian Authority and the Peace Process at Cross Roads', PALDEV Digest, no 2001-284 (1 November 2001).

13 Robert Fisk, 'Israel's black propaganda bid falters as documents reveal an impotent leader not a terrorist mastermind', *Independent* (9 May 2002).
14 Toufic Haddad, 'The age of no illusion', *Between the Lines*, vol. 2, no 17 (2002).
15 Malley and Agha, 'Camp David'.
16 Interview with Noah Kinnarty.
17 CIA, *The World Factbook*, 2000.
18 Frisch, *Countdown to Statehood*, p. 150.
19 Rouyer, *Turning Water into Politics*, pp. 150–1.
20 Feitelson, 'Implications of shifts in the Israeli water discourse for Israeli–Palestinian water negotiations', p. 305.
21 Saul Arlossoroff, 'The public commission on water sector reform', *International Water and Irrigation Review* (not dated). For discussion see Feitelson, 'Implications of shifts in the Israeli water discourse for Israeli–Palestinian water negotiations', p. 306.
22 Ibid.
23 Save the Children Fund West Bank and Gaza, 'Project Submission to DfID for Water and Sanitation Program in the Dura Area, Southern West Bank', p. 20.

Bibliography

Abbas, Mahmud, *Through Secret Channels* (Reading: Garnet, 1995).

Abd al-Shafi, Haider, 'The Oslo Agreement: an interview with Haider Abd al-Shafi', *Journal of Palestine Studies*, vol. 22, no 1 (1993), pp. 14–19.

Abdallah Saleh, Samir, 'The effects of Israeli occupation on the economy of the West Bank and Gaza Strip', in Jamal Nassar and Roger Heacock (eds), *Intifada: Palestine at the Crossroads* (New York: Praeger, 1991), pp. 37–51.

Abed, George (ed.), *The Palestinian Economy* (London: Routledge, 1988).

Abu-Amr, Ziad, 'Hamas: a historical and political background', *Journal of Palestine Studies*, vol. 22, no 4 (1993), pp. 5–19.

Abu-Amr, Ziad, 'The Palestinian Legislative Council: a critical assessment', *Journal of Palestine Studies*, vol. 26, no 4 (1997), pp. 90–7.

Abu-Lughod, Ibrahim, 'Introduction: on achieving independence', in Jamal Nassar and Roger Heacock (eds), *Intifada: Palestine at the Crossroads* (New York: Praeger, 1991).

Agha, Hussein, and Robert Malley, 'Camp David and after: a reply to Ehud Barak', *New York Review of Books* (13 June 2002).

Ahmad, Hisham, *Hamas: From Religious Salvation to Political Transformation: The Rise of Hamas in Palestinian Society* (Jerusalem: PASSIA, 1994).

Al-Alkim, Hassan Hamdan, *The GCC States in an Unstable World: Foreign Policy Dilemmas of Small States* (London: Saqi, 1994).

Alam, Undala, *Water Rationality: Mediating the Indus Treaty*, PhD Thesis (University of Durham, 1998).

Albin, Cecilia, *Justice and Fairness in International Negotiation* (Cambridge: Cambridge University Press, 2001).

Albin, Cecilia, 'When the weak confront the strong: justice, fairness and power in the Israeli–PLO Interim talks', *International Negotiation*, vol. 4, no 2 (1999), p. 339.

Aliewi, Amjad, and Ayman Jarrar, 'Technical Assessment of the Potentiality of the Herodian Wellfield against Additional Well Development Programmes', Report for the PWA (April 2000).

Aliewi, Amjad, and Anan Jayyousi, 'The Palestinian Water Resources in the Final Status Negotiations: Technical Framework and Professional Perception', Report for the PWA (4 May 2000).

Allan, Tony, 'Global Systems Ameliorate Local Droughts: Water, Food and Trade', SOAS Water Issues Group, Occasional Paper No 10 (1999).

Allan, Tony, *The Middle East Water Question: Hydropolitics and the Global Economy* (London: I.B.Tauris, 2000).

Allan, Tony, 'Striking the "right price" for water? Achieving harmony between basic human need, available resources and commercial viability', in Allan and Chibli Mallat (eds), *Water in the Middle East: Legal, Political and Commercial Implications* (London: I.B.Tauris, 1995).

Allan, Tony, 'Water in the Middle East and in Israel–Palestine: some local and global resource issues', in Marwan Haddad and Eran Feitelson, *Joint Management of Shared Aquifers: The Second Workshop* (Jerusalem: Palestine Consultancy Group and Harry S. Truman Institute for the Advancement of Peace, 1997).

Allan, Tony (ed.), *Water, Peace and the Middle East: Negotiating Resources in the Jordan Basin* (London: I.B.Tauris, 1996).

Allan, Tony, and Chibli Mallat (eds), *Water in the Middle East: Legal, Political and Commercial Implications* (London: I.B.Tauris, 1995).

Amery, Hussein, and Aaron Wolf (eds), *Water in the Middle East: A Geography of Peace* (Austin: University of Texas Press, 2000).

Amin, Samir, *Unequal Development* (Sussex: Harvester, 1976).

Amnesty International, 'Israel and the Occupied Territories. Demolition and Dispossession: The Destruction of Palestinian Homes' (December 1999).

Amnesty International, 'Israel and the Occupied Territories: The Heavy Price of Israeli Incursions' (14 April 2002).

Amnesty International, 'Palestinian Authority Silencing Dissent' (September 2000).

Anderson, Ewan, 'The political and strategic significance of water', *Outlook on Agriculture* (December 1992).

ARIJ, 'Environmental Profile for the West Bank', vol. 1: 'District of Bethlehem' (Bethlehem: ARIJ, June 1995).

ARIJ, 'Environmental Profile for the West Bank', vol. 3: 'Hebron District' (Bethlehem: ARIJ, November 1995).

Arlosoroff, Saul, 'The Israeli water law concept: summary', in Eran Feitelson and Marwan Haddad (eds), *Joint Management of Shared Aquifers: The Fourth Workshop* (Jerusalem: Palestine Consultancy Group and Harry S. Truman Institute for the Advancement of Peace, 1997), pp. 111–5.

Arlossoroff, Saul, 'The public commission on water sector reform', *International Water and Irrigation Review* (not dated).

Aronson, Geoffrey, 'Recapitulating the Agreements: The Israeli–PLO "Interim Agreements"', Centre for Policy Analysis on Palestine, Information Brief No 32 (27 April 2000).

Ashrawi, Hanan, *This Side of Peace* (New York: Simon and Schuster, 1995), esp. pp. 260–1.

Assaf, Karen, et al., *A Proposal for the Development of a Regional Water Master Plan* (Jerusalem: Israel/Palestine Centre for Research and Information, 1993).

Avineri, Shlomo, *The Making of Modern Zionism: The Intellectual Origins of the Jewish State* (London: Weidenfeld and Nicolson, 1981).

Bagis, Ali (ed.), *Water as an Element of Co-operation and Development in the Middle East* (Ankara: Hacettepe University and Friedrich Nauman Foundation, 1994).

Bailey, Sidney, *Four Arab–Israeli Wars and the Peace Process* (London: Macmillan, 1990).

Baran, Paul, *The Political Economy of Growth* (New York: Monthly Review, 1957).

Barlow, Maude and Tony Clarke, *Blue Gold: The Battle Against Corporate Theft of the World's Water* (London: Earthscan, 2002).

Bashara, Marwan, *Palestine/Israel: Peace or Apartheid* (London: Zed, 2001).

Baskin, Gershon, 'The West Bank and Israel's water crisis', in Baskin (ed.), *Water: Conflict or Co-operation?* (Jerusalem: Israel/Palestine Center for Research and Information, 1992).

Baumgarten, Helga, '"Discontented people" and "outside agitators": the PLO in the Palestinian uprising', in Jamal Nassar and Roger Heacock (eds), *Intifada: Palestine at the Crossroads* (New York: Praeger, 1991), pp. 207–26.

Beaumont, Peter, 'Conflict, coexistence, and cooperation: a study of water use in the Jordan basin', in Hussein Amery, and Aaron Wolf (eds), *Water in the Middle East: A Geography of Peace* (Austin: University of Texas Press, 2000).

Bellisari, A., 'Public health and the water crisis in the occupied Palestinian territories' *Journal of Palestine Studies*, vol. 23, no 2 (1994), pp. 52–63.

Ben Gurion University of the Negev and Tahal Consulting Engineers, 'Israel Water Study for the World Bank' (August 1994).

Ben-Eliezer, Uri, 'Is a military coup possible in Israel? Israel and French-Algeria in comparative historical-sociological perspective', *Theory and Society*, vol. 27 (1998), pp. 311–49.

Ben-Eliezer, Uri, *The Making of Israeli Militarism* (Bloomington: Indiana University Press, 1998).

Ben-Gai, T. et al., 'Long-term change in October rainfall patterns in southern Israel', *Theoretical and Applied Climatology*, vol. 46, no 4 (1993), pp. 209–17.

Ben-Gai, T. et al., 'Long-term changes in annual rainfall patterns in southern Israel', *Theoretical and Applied Climatology*, vol. 49, no 2 (1994), pp. 59–67.

Benn, Aluf, 'Import water from Turkey, prime minister's advisor urges', *Ha'aretz* (30 March 2000).

Benvenisti, Eyal, *Legal Dualism: The Absorption of the Occupied Territories into Israel* (Boulder: Westview, 1990).

Benvenisti, Meron, *Intimate Enemies: Jews and Arabs in a Shared Land* (Berkeley: University of California Press, 1995).

Benvenisti, Meron, *Report: Demographic, Economic, Legal, Social and Political Developments in the West Bank* (Boulder: Westview, 1986).

Benvenisti, Meron, *The West Bank Data Project: A Survey of Israel's Policies* (Washington DC: American Enterprise Institute, 1984).

Bergman, R., 'How much PA corruption is too much?' *Ha'aretz* (19 October 1999).

Bergman, R., 'Israel deposited NIS 1.5b in Arafat's personal account', *Ha'aretz* (8 October 1999).

Bill, James, and Robert Springborg, *Politics in the Middle East*, 4th edn. (New York: Harper Collins, 1994).

Biswas, Asit (ed.), *International Waters of the Middle East: From Euphrates-Tigris to Nile* (Bombay: Oxford University Press, 1994).

Boneh, Yohanan, and Uri Baida, 'Water sources in Judea and Samaria and their exploitation', in A. Shmueli et al. (eds), *Yehuda Veshomron* (Jerusalem: Kenaan, 1977–8), in Hebrew.

Brecher, Michael, *Decisions in Crisis: Israel, 1967 and 1973* (Berkeley: University of California Press, 1980).

Bromley, Simon, *Rethinking Middle East Politics: State Formation and Development* (Cambridge: Polity, 1994).

Bruck, Connie, 'The wounds of peace', *New Yorker* (14 October 1996).

Brynen, Rex, 'Buying peace? A critical assessment of international aid to the West Bank and Gaza', *Journal of Palestine Studies*, vol. 25, no 3 (1996), pp. 79–92.

Brynen, Rex, 'The dynamics of Palestinian elite formation', *Journal of Palestine Studies*, vol. 24, no 3 (1995), pp. 31–43.

Brynen, Rex, 'International aid to the West Bank and Gaza: a primer', *Journal of Palestine Studies*, vol. 25, no 2 (1996).

Brynen, Rex, 'The neopatrimonial dimension of Palestinian politics', *Journal of Palestine Studies*, vol. 25, no 1 (1995), pp. 23–36.

Brynen, Rex, *A Very Political Economy: Peacebuilding and Foreign Aid in the West Bank and Gaza* (Washington DC: US Institute for Peace, 2000).

Bulloch, John, and Adel Darwish, *Water Wars: Coming Conflicts in the Middle East* (London: Victor Gollancz, 1993).

Caponera, Dante, *Principles of Water Law and Administration, National and International* (Rotterdam: Balkimi, 1992).

Cardoso, Fernando, and Enzo Faletto, *Dependency and Development in Latin America* (Berkeley: University of California Press, 1979).

Carr, E.H., *The Twenty Years' Crisis, 1919–1939*, 2nd edn. (London: Macmillan, 1946).

CDM/Morganti, 'Task 4: Comprehensive Planning Framework for Palestinian Water Resources Development, Final Report, vol. 2', Report for the PWA (6 July 1997).

CDM/Morganti, 'Task 18: Study of the Sustainable Yield of the Eastern Aquifer Basin, Final Report', Report for the PWA (1998).

CDM/Morganti, 'Task 19: Two Stage Well Development Study for Additional Supplies in the West Bank, Stage 2', Report for the PWA (14 October 1997).

CDM/Morganti, 'Task 20: Monitoring Plan for the Eastern Aquifer Basin, Groundwater Monitoring Plan', Report for the PWA (10 September 1996).

CDM/Morganti, 'Task 25: Water Supply Facility Master Plan for the Hebron-Bethlehem Service Area, Final Report', Report for the PWA (8 March 1997).

CDM/Morganti, 'Task 26: Environmental Assessment for the Hebron-Bethlehem Water Supply System, Final Environmental Assessment', Report for the PWA (18 May 1997).

CDM/Morganti, 'Task 33: Integrated Water Resources Management Plan for the Hebron-Bethlehem Area, Final Plan', Report for the PWA (5 September 1997).

CDM/Morganti, *H_2O* (various).

Center for Palestine Research and Studies, 'Prevailing Perceptions of Aid Management', Research Report No 9 (December 1997).

Chomsky, Noam, *Fateful Triangle: The United States, Israel and the Palestinians*, 2nd edn. (London: Pluto, 1999).

Chomsky, Noam, 'Interview with Chomsky', *Z Magazine* (2 April 2002).

Chomsky, Noam, 'The Israel–Arafat agreement', *Z Magazine* (October 1993).

Chomsky, Noam, *World Orders, Old and New* (London: Pluto, 1994).

Christian Aid, 'Master or Servant: How Global Trade Can Work for the Benefit of Poor People' (2001).

CIA, *The World Factbook 2000* (2000).

Cippa, Andrea, 'Water Management in the City of Hebron', TIPH Information Report No 1945 (Hebron, October 1997).

Clarke, R., *Water: The International Crisis* (London: Earthscan, 1993).

Cohen, Amiram, 'Construction begins on imported water pipe', *Ha'aretz* (30 August 2000).

Collins, L., 'Up from the Palestinian sewage spews politics: sewage from the Autonomous areas is threatening Israel's main aquifers', *Jerusalem Post* (6 December 1996).

Collins, L., 'Water rights negotiations move along a very slippery road', *Jerusalem Post* (29 July 1995).

Cooley, John, 'The war over water', *Foreign Policy*, No 54 (1984), pp. 3–26.

Corbin, Jane, *Gaza First: The Secret Norway Channel to Peace between Israel and the PLO* (London: Bloomsbury, 1994).

Cordesman, Anthony, *Bahrain, Oman, Qatar and the UAE: Challenges of Security* (Boulder: Westview, 1997).

Cox, Robert, 'Social forces, states and world orders: beyond international relations theory', *Millennium*, vol. 10, no 2 (1981).

Dalton, Geraldine, 'Private Sector Finance for Water Infrastructure: What Does Cochabamba Tell Us About Using This Instrument?' SOAS Water Issues Group, Occasional Paper No 37 (2001).

Daniels, Glenda, 'Water privatisation test case "a total debacle"', *Mail and Guardian* (Johannesburg) (16 November 2001).

Daraghmeh, Mohammed, 'A scarred Nablus', *Palestine Report*, vol. 8, no 43 (24 April 2000).

Davidi, Asher, 'Israel's economic strategy for Palestinian independence', *Middle East Report*, no 184 (1993).

Davis, Uri, *Israel: An Apartheid State* (London: Zed, 1987).

Davis, Uri, Antonia Maks and John Richardson, 'Israel's water policies', *Journal of Palestine Studies*, vol. 9, no 2 (1980), pp. 3–31.

De Certeau, Michel, *The Practice of Everyday Life* (Berkeley: University of California Press, 1984).

De Jong, Jan, 'The geography of politics: Israel's settlement drive after Oslo', in George Giacaman and Dan Jorund Lonning (eds), *After Oslo: New Realities, Old Problems* (London: Pluto, 1998), p. 77.

Dean, Mitchell, *Governmentality: Power and Rule in Modern Society* (London: Sage, 1999).

Deleuze, Gilles, and Felix Guattari, *A Thousand Plateaus: Capitalism and Schizophrenia* (London: Athlone, 1987).

Dellapenna, Joseph, 'Developing a treaty regime for the Jordan Valley', in Eran Feitelson and Marwan Haddad (eds), *Joint Management of Shared Aquifers: The Fourth Workshop* (Jerusalem: Harry S. Truman Institute for the Advancement of Peace, and Palestine Consultancy Group, 1995).

Di Giovanni, Janine, 'Children scream for water in the "city of bombers"', *The Times* (London) (9 April 2002).

Dillman, Jeffrey, 'Water rights in the Occupied Territories', *Journal of Palestine Studies*, vol. 19, no 1 (1989), pp. 46–71.

Dolatyar, Mustapha, and Tim Gray, 'The politics of water scarcity in the Middle East', Environmental Politics, vol. 9, no 3 (2000).

Dolatyar, Mustapha, and Tim Gray, *Water Politics in the Middle East: A Context for Conflict or Co-operation?* (London: Macmillan, 2000).

Doyle, Michael, 'Kant, liberal legacies, and foreign affairs: parts 1 and 2', *Philosophy and Public Affairs*, vol. 12 (1983), pp. 205–34 and 323–53.

Dupuy, Trevor, *Elusive Victory: The Arab–Israeli Wars, 1947–1974* (London: Macdonald and Jane's, 1978).

Efrat, Ron, 'Interview with Adel Samara: the hidden logic of Arafat's surrender', *Challenge* (June 2000).

Efrat, Ron, 'Porcupine tangos: the PA and the NGOs (accompanied by a CIA orchestra)', *Challenge*, no 57.

Ehrlich, Paul, *The Population Bomb* (New York: Ballantine, 1968).

Ehrlich, Paul, and Anne Ehrlich, *The Population Explosion* (New York: Touchstone, 1991).

Eisenstadt, Shmuel, *Israeli Society* (London: Weidenfeld and Nicolson, 1967).

Eisenstadt, Shmuel, *Traditional Patrimonialism and Modern Neo-Patrimonialism* (London: Sage, 1973).

Eldar, Akiva, 'Moratinos Document' reprinted with commentary in *Ha'aretz* (14 February 2002).

El-Hindi, Jamal, 'Note: the West Bank aquifer and conventions regarding laws of belligerent occupation', *Michigan Journal of International Law*, vol. 11, no 4 (1990), pp. 1400–23.

Elmusa, Sharif, *Negotiating Water: Israel and the Palestinians* (Washington DC: Institute for Palestine Studies, 1996).

Elmusa, Sharif, *Water Conflict: Economics, Politics, Law and the Palestinian–Israeli Water Resources* (Washington DC: Institute for Palestine Studies, 1997).

Elmusa, Sharif, and Mahmud El-Jaafari, 'Power and trade: the Israeli–Palestinian economic protocol', *Journal of Palestine Studies*, vol. 24, no 2 (1995), pp. 14–32.

E-WaSH, 'Nablus water situation', Internal Report (14 April 2002).

E-WaSH, 'Ramallah water situation', Internal Report (14 April 2002).

E-WaSH, 'Tulkarm water situation', Internal Report (14 April 2002).

Falkenmark, Malin, 'Fresh water: time for a modified approach', *Ambio*, vol. 15, no 4 (1986).

Falkenmark, Malin, 'The massive water scarcity now threatening Africa: why isn't it being addressed?' *Ambio*, vol. 18, no 2 (1989), pp. 112–8.

Falkenmark, Malin, 'Middle East hydropolitics: water scarcity and conflicts in the Middle East', *Ambio*, vol. 18, no 6 (1989), pp. 350–2.

Falkenmark, Malin, 'Vulnerability generated by water scarcity', *Ambio*, vol. 18, no 6 (1989), pp. 352–3.

Falkenmark, Malin, and Jan Lundqvist, 'Looming water crisis: new approaches are inevitable', in Leif Ohlsson (ed.), *Hydropolitics: Conflicts Over Water as a Development Constraint* (London: Zed, 1995).

Falkenmark, Malin, and Jan Lundqvist, 'Towards water security: political determination and human adaptation crucial', *Natural Resources Forum*, vol. 21, no 1 (1998).

Falkenmark, Malin, and Carl Widstrand, 'Population and Water Resources: A Delicate Balance', *Population Bulletin*, vol. 47, no 3 (1992).

Farsoun, Samih, and Jean Landis, 'The sociology of an uprising: the roots of the intifada', in Jamal Nassar and Roger Heacock (eds), *Intifada: Palestine at the Crossroads* (New York: Praeger, 1991), pp. 15–35.

Feitelson, Eran, 'The ebb and flow of the Arab–Israeli water conflict: are past confrontations likely to resurface?' *Water Policy*, vol. 2 (2000).

Feitelson, Eran, 'Implications of shifts in the Israeli water discourse for Israeli–Palestinian water negotiations', *Political Geography*, vol. 21 (2002).

Feitelson, Eran, and Marwan Haddad (eds), *Joint Management of Shared Aquifers: The First Workshop* (Jerusalem: Harry S. Truman Institute for the Advancement of Peace, and Palestine Consultancy Group, 1994).

Feitelson, Eran, and Marwan Haddad (eds), *Joint Management of Shared Aquifers: Final Report* (Jerusalem: Harry S. Truman Institute for the Advancement of Peace, and Palestine Consultancy Group, 1995).

Feitelson, Eran, and Marwan Haddad (eds), *Joint Management of Shared Aquifers: The Fourth Workshop* (Jerusalem: Harry S. Truman Institute for the Advancement of Peace, and Palestine Consultancy Group, 1998).

Ferguson, James, *The Anti-Politics Machine: 'Development', Depoliticization and Bureaucratic Power in Lesotho* (Cambridge: Cambridge University Press, 1990).

Fetter, Charles, *Applied Hydrogeology*, 3rd edn. (New Jersey: Prentice Hall, 1994).

Finkelstein, Norman, *Image and Reality of the Israel–Palestine Conflict* (London: Verso, 1995).

Finkelstein, Norman, 'The real meaning of the Wye River Memorandum' (28 July 1998).

Finkelstein, Norman, 'Whither the peace process', *New Left Review*, no 218 (July/August 1996).

Fisk, Robert, 'Israel's black propaganda bid falters as documents reveal an impotent leader not a terrorist mastermind', *Independent* (9 May 2002).

Flapan, Simha, *The Birth of Israel: Myths and Realities* (London: Croom Helm, 1987).

Fletcher, Elaine, 'Israel, PLO Make Deal on West Bank Water', *The San Francisco Examiner* (21 September 1995).

'Flowing uphill', *The Economist* (12 August 1995).

Foucault, Michel, 'Afterword: the subject and power', in Hubert Dreyfus and Paul Rabinow, *Michel Foucault: Beyond Structuralism and Hermeneutics* (Hemel Hempstead: Harvester Wheatsheaf, 1982).

Foucault, Michel, *Discipline and Punish: The Birth of the Prison* (London: Penguin, 1977).

Foucault, Michel, 'Governmentality', *I & C*, no 4 (1979), pp. 5–21.

Foucault, Michel, *Politics, Philosophy, Culture: Interviews and Other Writings, 1977–1984*, ed. Lawrence Kritzman (London: Routledge, 1988).

Foucault, Michel, *Power/Knowledge: Selected Interviews and Other Writings 1972–1977* (Brighton: Harvester, 1980).

Foundation for Middle East Peace, *Report on Israeli Settlement* (various).

Frank, Andre Gunther, *Capitalism and Underdevelopment in Latin America* (London: Monthly Review, 1967).

Frey, Frederick, and Thomas Naff, 'Water: an emerging issue in the Middle East?' *Annals of the American Academy of Political and Social Science*, vol. 482 (1985).

Frisch, Hillel, *Countdown to Statehood: Palestinian State Formation in the West Bank and Gaza* (Albany: State University of New York Press, 1998).

Galnoor, Itzhak, 'Water Planning: Who Gets the Last Drop?' in R. Bilski (ed.), *Can Planning Replace Politics? The Israeli Experience* (The Hague: Martinus Nijhoff 1980), pp. 137–215.

Galnoor, Itzhak, 'Water Policymaking in Israel', in Hillel Shuval (ed.), *Water Quality Management Under Conditions of Scarcity: Israel as a Case Study* (New York: Academic Press 1980) pp. 287–314.

Garfinkle, Adam, *Israel and Jordan in the Shadow of War: Functional Ties and Futile Diplomacy in a Small Place* (London: Macmillan, 1992).

Garfinkle, Adam, *War, Water and Negotiation in the Middle East: The Case of the Palestine–Syria Border, 1916–1923* (Tel Aviv: Moshe Dayan Center, Tel Aviv University, 1994).

Gazit, Shlomo, *The Carrot and the Stick: Israeli Command of the West Bank and Gaza* (Nicosia: Beisan, 1985), pp. 91–2.

Gellman, Barton, 'Allied air war struck broadly in Iraq; officials acknowledge strategy went beyond purely military targets', *Washington Post* (23 June 1991).

Gellner, Ernest, *Muslim Society* (Cambridge: Cambridge University Press, 1981).

Gerges, Fawaz, *The Superpowers and the Middle East: Regional and International Politics, 1955–1967* (Boulder: Westview, 1994).

Gerner, Deborah, and Philip Schrodt, 'Middle East politics,' in Gerner (ed.), *Understanding the Contemporary Middle East* (Boulder: Lynne Rienner, 2000).

Giacaman, George, 'In the throes of Oslo: Palestinian society, civil society and the future', in Giacaman and Dan Jorund Lonning (eds), *After Oslo: New Realities, Old Problems* (London: Pluto, 1998), pp. 1–15.

Giacaman, Rita, *Life and Health in Three Palestinian Villages* (London: Ithaca, 1988).

Giddens, Anthony, *A Contemporary Critique of Historical Materialism*, vol. II: *The Nation-State and Violence* (Cambridge: Polity, 1985).

Gilpin, Robert, *War and Change in World Politics* (New York: Cambridge University Press, 1981).

Gleick, Peter, 'The implications of global climatic changes for international security', *Climatic Change*, vol. 15 (1989), pp. 309–25.

Gleick, Peter, *Water in Crisis: A Guide to the World's Fresh Water Resources* (New York: Oxford University Press, 1993).

Gleick, Peter, 'Water, war and peace in the Middle East', *Environment*, vol. 36, no 3 (1994), pp. 6–15, 35–42.

'Government to create water desalting facility', *Ha'aretz* (18 April 2000).

Government of Israel, 'Palestinian Obligations as Per Note for the Record of the Hebron Protocol of January 15, 1997' (13 January 1998).

Gresh, Alain, *The PLO: The Struggle Within* (London: Zed, 1988).

Grieco, Joseph, *Cooperation Among Nations: Europe, America and Non-Tariff Barriers to Trade* (London: Cornell University Press, 1990).

Groth, Allon, *The PLO's Road to Peace: Processes of Decision-Making* (London: Royal United Services Institute for Defence Studies, 1995).

Gruen, George, *The Water Crisis: The Next Middle East Crisis?* (Los Angeles: Wiesenthal Center, 1991).

GTZ, 'Middle East Regional Study on Water Supply and Demand Development, Concluding Report' (February 1998).

GTZ, 'Middle East Regional Study on Water Supply and Demand Development, Phase 1 Report' (August 1996).

Guttman, Yossi, 'Hydrogeology of the Eastern Aquifer in the Judea Hills and the Jordan Valley', Report for the German–Israeli–Jordanian–Palestinian Joint Research Program for the Sustainable Utilization of Aquifer Systems (18 January 1998).

Guyatt, Nick, *The Absence of Peace: Understanding the Israeli–Palestinian Conflict* (London: Zed, 1988).

Haas, Peter (ed.), 'Knowledge, power and international policy coordination', *International Organization* (special issue), vol. 46, no 1 (1992).

Haas, Peter, *Saving the Mediterranean: The Politics of International Environmental Cooperation* (New York: Columbia University Press, 1990).

Haddad, Marwan, and Eran Feitelson, *Joint Management of Shared Aquifers: The Second Workshop* (Jerusalem: Palestine Consultancy Group and Harry S. Truman Institute for the Advancement of Peace, 1997).

Haddad, Marwan, and Eran Feitelson (eds), *Joint Management of Shared Aquifers: The Third Workshop* (Jerusalem: Palestine Consultancy Group and Harry S. Truman Institute for the Advancement of Peace, 1997).

Haddad, Toufic, 'The age of no illusion', *Between the Lines*, vol. 2, no 17 (2002).

Haidar, Aziz, *On the Margins: The Arab Population in the Israeli Economy* (London: Hurst and Co, 1995).

Halevi, Ilan, 'Self-government, democracy and mismanagement under the Palestinian Authority', *Journal of Palestine Studies*, vol. 27, no 3 (1998), pp. 35–48.

Halliday, Fred, and Justin Rosenberg, 'Interview with Ken Waltz', *Review of International Studies*, vol. 24 (1998), pp. 371–86.

Hamzeh-Muhaisen, M., 'Who's to blame for the severe water shortages?' *Palestine Report*, vol. 5, no 9 (14 August 1998).

Harel, Amos, 'IDF admits "ugly vandalism" against Palestinian property', *Ha'aretz* (30 April 2002).

Harvey, David, *Justice, Nature and the Geography of Difference* (Oxford: Blackwell, 1996).

Hass, Amira, 'Chairman Arafat straightens out his financial accounts', *Ha'aretz* (13 January 2000).

Hass, Amira, 'Cut and dried', *Ha'aretz* (31 July 1998).

Hass, Amira, 'Dire water shortages in West Bank', *Ha'aretz* (27 August 1998).

Hass, Amira, 'A report on Palestinian water crisis', *Al Quds* (25 June 1999).

Hass, Amira, 'Sharon says PA excuses are all wet', *Ha'aretz* (19 August 1998).

Hass, Amira, '25,000 lack water in Ramallah', *Ha'aretz* (2 April 2002).

Hellman, Z., and P. Inbari, 'Water Pollution in the West Bank: Overcoming Political Stumbling Blocks to a Solution', Institute for Peace Implementation (29 April 1997).

Herbert, Frank, *Dune* (London: Hodder and Stoughton, 1965).

Hilal, Jamil, 'The effect of the Oslo Agreement on the Palestinian political system', in George Giacaman and Dan Jorund Lonning (eds), *After Oslo: New Realities, Old Problems* (London: Pluto, 1998), pp. 121–45.

Hillel, Daniel, *Rivers of Eden: The Struggle for Water and the Quest for Peace in the Middle East* (New York: Oxford University Press, 1994).

Hilterman, Joost, *Behind the Intifada: Labour and Women's Movements in the Occupied Territories* (Princeton: Princeton University Press, 1991).

Hirsch, M., 'Game theory, international law and future environmental co-operation in the Middle East', *Denver Journal of International Law and Policy*, vol. 27 (1998), pp. 75–119.

Hoffman, D., 'Shamir plan was to stall autonomy: Rabin says he'll cut subsidies to settlers', *Washington Post* (27 June 1992).

Honig-Parnass, Tigva, 'A new stage: military intifada, Israeli panic, ruthless repression', *News From Within* (2 April 1993) pp. 1–4.

Horkheimer, Max, *Critical Theory: Selected Essays* (New York: Herder and Herder, 1972).

Hurewitz, J., *Diplomacy in the Near and Middle East: A Documentary Record: 1914–1956*, vol. 2 (New York: Octagon, 1956).

ICRC, 'Iraq: A Decade of Sanctions' (14 December 1999).

IHS, 'The Development, Exploitation and Condition of Groundwater Sources in Israel up to the Fall of 1997' (1998).

Integrated Social Development Centre, 'Water Privatisation in Ghana: An Analysis of Government and World Bank Policies' (Accra: ISDC, 2001).

Isaac, Jad, 'Core issues of the Palestinian–Israeli water dispute', in K. Spillman and G. Bachler (eds), *Environmental Crisis: Regional Conflicts and Ways of Cooperation*, Environment and Conflicts Project, Occasional Paper No 14 (Zurich: Centre for Security Studies and Conflict Research, 1995).

Isaac, Jad 'A Palestinian perspective on the water crisis', *Israel–Palestine Journal of Politics, Economics and Culture*, vol. 5, no 1 (1998), pp. 54–8.

Isaac, Jad, 'Water and Palestinian–Israeli peace negotiations', *Palestine Center: News and Analysis* (19 August 1999).

Isaac, Jad, and Jan Selby, 'The Palestinian water crisis: status, projections and potential for resolution', *Natural Resources Forum*, vol. 20 (1996), pp. 18–20.

Isaac, Jad, and Hillel Shuval (eds), *Water and Peace in the Middle East* (Amsterdam: Elsevier, 1994).

Israel and the PLO, 'Agreement on the Gaza Strip and the Jericho Area' (Cairo, 4 May 1994).

Israel and the PLO, 'Agreement on Preparatory Transfer of Powers and Responsibilities' (Erez, 19 August 1994).

Israel and the PLO, 'Declaration of Principles on Interim Self-Government Arrangements' (Washington DC, 13 September 1993).

Israel and the PLO, 'Interim Agreement on the West Bank and Gaza Strip' (Washington DC, 28 September 1995).

Israel and the PLO, 'Protocol Concerning the Redeployment in Hebron' (Jerusalem, 17 January 1997).

Israel and the PLO, 'Protocol on Economic Relations' (Paris, 29 April 1994).

Israel and the PLO, 'Sharm-el-Sheikh Memorandum on Implementation of Outstanding Commitments of Agreements Signed and the Resumption of Permanent Status Negotiations' (Sharm-el-Sheikh, 4 September 1999).

Israel and the PLO, 'Wye River Memorandum' (Washington DC, 23 October 1998).

Israeli Central Bureau of Statistics, 'Monthly Bulletin of Statistics', vol. 51 (March 2000).

Israeli–Palestinian Joint Water Committee, 'Joint Declaration for Keeping the Water Infrastructure Out of the Cycle of Violence' (Erez, 31 January 2001).

Jansen, Michael, 'The peace process flounders in Geneva', *Middle East International* (7 April 2000).

Jehl, D., 'Water divided haves from have-nots in West Bank', *New York Times* (15 August 1998).

Jerusalem Media and Communication Centre, *Water: The Red Line* (Jerusalem: JMCC, 1994).

JMCC, 'Foreign Aid and Development in Palestine, Phase I Report' (1997).

JMCC, 'Foreign Aid and Development in Palestine, Phase III Report' (1999).

JMCC, 'Mortgaging Self-Reliance: Foreign Aid and Development in Palestine, Phase II Report' (1997).

Jones, J. Anthony, *Global Hydrology: Processes, Resources and Environment* (Harlow: Longman, 1997).

Jordanian National Information System, Department of Statistics, 'Population Projection 1998–2005' (1998).

Kally, Elisha, and Gideon Fishelson, *Water and Peace: Water Resources and the Arab–Israeli Peace Process* (Westport: Praeger, 1993).

Kanarek, A., and M. Michail, 'Groundwater recharge with municipal effluent: Dan Region Reclamation Project, Israel', *Water Science and Technology*, vol. 34, no 11 (1996), pp. 227–33.

Kant, Immanuel, *Kant's Political Writings*, ed. H. Reiss (Cambridge: Cambridge University Press, 1970).

Kartin, Amnon, 'Factors inhibiting structural changes in Israel's water policy', *Political Geography*, vol. 19 (2000).

Keohane, Robert, *After Hegemony: Co-operation and Discord in the World Political Economy* (Princeton: Princeton University Press, 1984).

Keohane, Robert, 'The theory of hegemonic stability and changes in international economic regimes, 1967–1977', in Ole Holsti et al. (eds), *Change in the International System* (Boulder: Westview, 1980), pp. 131–62.

Khalidi, Ahmad, 'Security in the final Middle East settlement: some components of Palestinian national security', *International Affairs*, vol. 71, no 1 (1995), pp. 1–18.

Khalidi, Walid, 'The Gulf crisis: origins and consequences', *Journal of Palestine Studies*, vol. 20, no 2 (1991), pp. 5–28.

Kidron, Peretz, 'A bomb hardens the mood', *Middle East International* (25 August 1995).

Kimmerling, Baruch, 'Boundaries and frontiers of the Israeli control system', in Kimmerling (ed.), *The Israeli State and Society: Boundaries and Frontiers* (Albany: State University of New York Press, 1989).

Kimmerling, Baruch, *Zionism and Territory: The Socio-Territorial Dimension of Zionist Politics* (Berkeley: University of California Press, 1983).

King, John, *Handshake in Washington: The Beginning of Middle East Peace?* (London: Ithaca, 1994).

King, L., 'Mideast neighbors quarrel over water', *Associated Press* (7 August 1999).

Kinnarty, Noah, 'An Israeli view: if only they were quiet, the Palestinians have numerous opportunities', Interview, *Bitterlemons* (5 August 2002).

Kleiman, Ephraim, 'The place of manufacturing in the growth of the Israeli economy', *Journal of Development Studies*, vol. 3, no 3 (1967).

Kleiman, Ephraim, 'The waning of Israeli etatism', *Israel Studies*, vol. 2, no 2 (1988), pp. 146–71.

Kliot, Nurit, 'A cooperative framework for sharing scarce water resources: Israel, Jordan, and the Palestinian Authority', in Hussein Amery and Aaron Wolf (eds), *Water in the Middle East: A Geography of Peace* (Austin: University of Texas Press, 2000).

Kliot, Nurit, *Water Resources and Conflict in the Middle East* (London: Routledge, 1994).

Kra, Baruch, 'Settlement building up 81% in first quarter', *Ha'aretz* (22 August 2000).

Kronfeld, J. et al., 'Natural isotopes and water stratification in the Judea Group aquifer (Judean Desert)', *Israel Journal of Earth Sciences*, vol. 39 (1992), pp. 71–6.

Laqueur, Walter, *The Road to War: The Origins and Aftermath of the Arab–Israeli Conflict, 1967–8* (London: Weidenfeld and Nicolson, 1968).

Laqueur, Walter, and Barry Rubin (eds), *The Israel–Arab Reader: A Documentary History of the Middle East Conflict* (New York: Penguin, 1984).

Latour, Bruno, *Science in Action: How to Follow Scientists and Engineers Through Society* (Cambridge MA: Harvard University Press, 1987).

Latour, Bruno, *We Have Never Been Modern* (Cambridge, MA: Harvard University Press, 1993).

LAW, 'Bypass Road Construction in the West Bank: The End of the Dream of Palestinian Sovereignty', Jerusalem (1996).

Lees, Sarah, *The Political Ecology of the Water Crisis in Israel* (Lanham: University Press of America, 1998).

Levy, Gideon, 'The sewage of Ma'aleh Adumin', *Ha'aretz* (22 February 1998).

Libiszewski, S., 'Water Disputes in the Jordan Basin Region and their Role in the Resolution of the Arab–Israeli Conflict', Environment and Conflicts

Project, Occasional Paper No 13 (Zurich: Center for Security Studies and Conflict Research, 1995).

Lindholm, Helena, 'Water and the Arab–Israeli conflict', in Leif Ohlsson (ed.), *Hydropolitics: Conflicts Over Water as a Development Constraint* (London: Zed, 1995).

Local Aid Co-ordination Committee Co-Chairs, 'Damage to Civilian Infrastructure and Institutions in the West Bank estimated at US$361 million', Press Release (15 May 2002).

Lockman, Zachary, and Joel Beinin (eds), *Intifada: The Palestinian Uprising Against Israeli Occupation* (Washington DC: MERIP, 1989).

Lonergan, Steve, and David Brooks, *Watershed: The Role of Fresh Water in the Israeli–Palestinian Conflict* (Ottawa: IDRC, 1994).

Lowi, Miriam, *Water and Power: The Politics of a Scarce Resource in the Jordan Basin Area*, 2nd edn. (Cambridge: Cambridge University Press, 1995).

Lustick, Ian, *Arabs in the Jewish State: Israel's Control of a National Minority* (Austin: University of Texas Press, 1980).

Makovsky, David, 'Oslo is not Dead. It Cannot Die', *Ha'aretz* (28 August 1998).

Mallat, Chibli, 'The quest for water use principles: reflections on Shari'a and custom in the Middle East', in Tony Allan and Chibli Mallat (eds), *Water in the Middle East: Legal, Political and Commercial Implications* (London: I.B.Tauris, 1995), pp. 127–37.

Malley, Robert, and Hussein Agha, 'Camp David: the tragedy of errors', *New York Review of Books* (9 August 2001).

Malthus, Thomas, *An Essay on the Principle of Population* (Cambridge: Cambridge University Press, 1989).

Marx, Karl, *Capital*, vol. I (London: Pelican, 1976).

Marx, Karl, *Early Writings*, ed. T.B. Bottomore (London: Watts and Co, 1963).

Marx, Karl, 'Preface to "A Contribution to the Critique of Political Economy"', in Marx and Engels, *Selected Works*, vol. 1 (Moscow: Foreign Languages Publishing, 1962).

Marx, Karl, 'Theses on Feuerbach', in *Early Writings*, ed. Lucio Colletti (London: Penguin, 1975).

Marx, Marl, and Friedrich Engels, *The German Ideology* (London: Lawrence and Wishart, 1965).

Masalha, Nur, 'A critique on Benny Morris', *Journal of Palestine Studies*, vol. 21, no 1 (1991), pp. 90–7.

Masalha, Nur, *Expulsion of the Palestinians: The Concept of Transfer in Zionist Political Thought, 1882–1948* (Washington: Institute for Palestine Studies, 1992).

Mazor, Emanuel, and Magda Molcho, 'Geochemical studies on the Feshcha Springs, Dead Sea basin', *Journal of Hydrology*, vol. 15 (1971), pp. 37–47.

McCaffrey, Stephen, 'Water, politics and international law', in Peter Gleick, *Water in Crisis: A Guide to the World's Fresh Water Resources* (New York: Oxford University Press, 1993), pp. 92–104.

McDowall, David, *The Palestinians: The Road to Nationhood* (London: Minority Rights Publications, 1994).

McKibben, Bill, *The End of Nature* (New York: Viking, 1990).

McNamara, Robert, *The Essence of Security: Reflections in Office* (London: Hodder and Stoughton, 1968).

Meadows, Donella et al., *The Limits to Growth: A Report for the Club of Rome's Project on the Predicament of Mankind* (New York: Earth Island, 1972).

Meinzer, Oscar, 'Outline of Groundwater Hydrology with Definitions', US Geological Survey, Water Supply Paper 494 (1923).

Migdal, Joel, *Strong Societies and Weak States: State-Society Relations and State Capabilities in the Third World* (Princeton: Princeton University Press, 1988).

Migdal, Joel, *Through the Lens of Israel: Explorations in State and Society* (Albany: State University of New York Press, 2001).

Mills, C. Wright, *The Sociological Imagination* (Oxford: Oxford University Press, 1959).

Mitrany, David, *The Functional Theory of Politics* (New York: St. Martin's Press, 1975).

Mitrany, David, *A Working Peace System* (London: Royal Institute for International Affairs, 1943).

Monbiot, George, 'They're all damned', *The Guardian* (26 February 2002).

Moore, James, 'An Israeli–Palestinian water sharing regime', in Jad Isaac and Hillel Shuval (eds), *Water and Peace in the Middle East* (Amsterdam: Elsevier, 1994), pp. 181–92.

Moore, James, 'Parting the waters: calculating Israeli and Palestinian entitlements to the West Bank aquifers and the Jordan River basin', *Middle East Policy*, vol. 3, no 1 (1993), pp. 91–108.

MOPIC, 'Regional Plan for West Bank Governorates: Water and Wastewater' (July 1998).

Morgenthau, Hans, *Politics Among Nations: The Struggle for Power and Peace*, 6th edn. (New York: Knopf, 1985).

Morris, Benny, *The Birth of the Palestinian Refugee Problem, 1947–1949* (Cambridge: Cambridge University Press, 1987).

Morris, Benny, 'Camp David and after: an interview with Ehud Barak', *New York Review of Books* (13 June 2002).

Morris, Benny, Ehud Barak, Hussein Agha and Robert Malley, 'Camp David and after – continued', *New York Review of Books* (27 June 2002).

Murakami, Masahiro, *Managing Water for Peace in the Middle East: Alternative Strategies* (Tokyo: United Nations University Press, 1995).

Murphy, Emma, 'Israel and the Palestinians: The Economic Rewards of Peace', CMEIS Occasional Paper No 47 (March 1995).

Murphy, Emma, 'Stacking the deck: the economics of the Israeli–PLO accords', *Middle East Report*, vol. 25, no 3/4 (1995), pp. 35–8.

Naff, Thomas, and Ruth Matson (eds), *Water in the Middle East: Conflict or Cooperation?* (Boulder: Westview, 1984).

Neff, Donald, 'The US piles on the pressure', *Middle East International*, no 631 (18 August 2000).
Neff, Donald, *Warriors for Jerusalem: The Six Days that Changed the Middle East* (New York: Simon and Schuster, 1984).
'Negotiators achieve breakthrough on water rights', *Israel-Line* (25 August 1995).
Newman, David, *The Impact of Gush Enumin: Politics and Settlement in the West Bank* (London: Croom Helm, 1985).
Ohlsson, Leif, *Environment, Scarcity and Conflict: A Study of Malthusian Concerns*, Department of Peace and Conflict Research, Goteburg University (1999).
Ohlsson, Leif (ed.), *Hydropolitics: Conflicts Over Water as a Development Constraint* (London: Zed, 1995).
Owen, Roger, *State, Power and Politics in the Making of the Modern Middle East* (London: Routledge, 1992).
Oxfam, 'An Urgent Call to Address the Human Costs of the Israeli–Palestinian Conflict', Oxfam Briefing Note (4 April 2002).
Oxfam, 'Crisis in southern Africa', Briefing Paper 23 (June 2002).
'PA cracks down on debtors', *Middle East Newsline* (26 September 2000).
PA, 'Law No. 2 for 1996: Concerning the Establishment of the PWA' (issued by President Arafat, 18 January 1996).
Pacheco, Allegra, 'Oslo II and Still No Water', *Middle East International* (3 November 1995), pp. 18–19.
Palestinian Central Bureau of Statistics, 'The Palestinian Census of Population, Housing and Establishments: Preliminary Results' (December 1997).
Palestinian NGO Emergency Initiative in Jerusalem, 'Report on the Destruction to Palestinian Governmental Institutions in Ramallah Caused by IDF Forces Between March 29 and April 21, 2002' (22 April 2002).
Palestinian NGO Emergency Initiative in Jerusalem, 'Destruction of Non-Governmental Organizations in Ramallah Caused by IDF Forces Between March 29 and April 21, 2002' (22 April 2002).
Pappe, Ilan, *The Making of the Arab–Israeli Conflict, 1947–1951* (London: I.B.Tauris, 1992).
Parker, Christopher, *Resignation or Revolt: Socio-Political Development and the Challenges of Peace in Palestine* (London: I.B.Tauris, 1999).
Parker, Richard, *The Politics of Miscalculation in the Middle East* (Bloomington: Indiana University Press, 1993).
Parsons, Talcott, *The Structure of Social Action* (New York: Free Press, 1949).
Pearce, Fred, *The Damned: Rivers, Dams and the Coming World Water Crisis* (London: Bodley Head, 1992).
Peled, Yoav, 'From Zionism to capitalism: the political economy of Israel's decolonization of the occupied territories', *Middle East Report*, no 194/5 (1995), pp. 13–17.
Peretz, Don, *The West Bank: History, Politics, Society and Economy* (Boulder: Westview, 1986).

Peters, Joel, *Pathways to Peace: The Multilateral Arab–Israeli Peace Talks* (London: Royal Institute for International Affairs, 1996).
Peterson, S., 'What could float – or sink – peacemaking', *Christian Science Monitor* (July 14 1999).
PHG, 'Water for Palestine: A Critical Assessment of the European Investment Bank's Lending Strategy in the Rehabilitation of Water Resources in the Southern West Bank', Report for the Reform the World Bank Campaign (2000).
Pope, Hugh, 'Water in a bag', *Middle East International*, no 377 (8 June 1990), p. 14.
Porrit, Jonathan, *Seeing Green: The Politics of Ecology Explained* (Oxford: Basil Blackwell, 1984).
PWA, 'West Bank Water Facilities Map' (1996).
Quandt, William, *American Diplomacy and the Arab–Israeli Conflict Since 1967* (Berkeley: University of California Press, 1993).
Reeves, Phil, 'New barriers widen gulf on West Bank', *The Independent* (26 May 2002).
'Report of the Sharm El-Sheikh Fact-Finding Committee', chaired by former US Senator George Mitchell (30 April 2001).
Rich, Bruce, *Mortgaging the Earth: The World Bank, Environmental Impoverishment and the Crisis of Development* (London: Earthscan, 1994).
Rigby, Andrew, *Living the Intifada* (London: Zed, 1991).
Robinson, Glen, *Building a Palestinian State: The Incomplete Revolution* (Bloomington: Indiana University Press, 1997).
Robinson, Glen, 'The growing authoritarianism of Arafat's regime', *Survival*, vol. 39, no 2 (1997), pp. 42–56.
Rodinson, Maxime, *Israel: A Colonial-Settler State?* (New York: Monad, 1973).
Rofe and Raffety Consulting Engineers, 'Jerusalem and District Water Supply, Geological and Hydrological Report', Report for the Central Water Authority, Hashemite Kingdom of Jordan (1963).
Rofe and Raffety, 'Nablus District Water Resources Survey, Geological and Hydrological Report', Report for the Central Water Authority, Hashemite Kingdom of Jordan (1965).
Rofe and Raffety, 'West Bank Hydrology, 1963–65', Report for the Central Water Authority, Hashemite Kingdom of Jordan (1965).
Rogers, Peter, and Peter Lydon (eds), *Water in the Arab World: Perspectives and Prognoses* (Cambridge MA: Harvard University Press, 1994).
Rosencrance, Richard, *The Rise of the Trading State* (New York: Basic Books, 1986).
Rostow, Walter, *The Stages of Economic Growth: A Non-Communist Manifesto* (Cambridge: Cambridge University Press, 1960).
Rouyer, Alwyn, 'Implementation of the water accords in the Oslo II Agreement', *Middle East Policy*, vol. 7 (1999).
Rouyer, Alwyn, *Turning Water into Politics: The Water Issue in the Palestinian–Israeli Conflict* (London: Macmillan, 2000).

Rouyer, Alwyn, 'The water issue in the Israeli–Palestinian peace process', *Survival*, vol. 39, no 2 (1997), pp. 57–81.

Rouyer, Alwyn, 'Zionism and water: influences on Israel's future water policy during the pre-state period', *Arab Studies Quarterly*, vol. 18, no 4 (1996), pp. 25–47.

Rowley, Gwyn, 'The West Bank: native water resource systems and competition', *Political Geography Quarterly*, vol. 9, no 1 (1990), pp. 39–52.

Roy, Sara, 'De-development revisited: Palestinian economy and society since Oslo', *Journal of Palestine Studies*, vol. 28, no 3 (1999), pp. 64–82.

Roy, Sara, 'Gaza: new dynamics of civil disintegration', *Journal of Palestine Studies*, vol. 22, no 4 (1993), pp. 22–37.

Roy, Sara, *The Gaza Strip: The Political Economy of De-Development* (Washington DC: Institute for Palestine Studies, 1995).

Roy, Sara, *The Palestinian Economy and the Oslo Process: Decline and Fragmentation* (Abu Dhabi: Emirates Center for Strategic Studies and Research, 1998).

Roy, Sara, 'Palestinian society and economy: the continued denial of possibility', *Journal of Palestine Studies*, vol. 30, no 4 (2001), pp. 5–20.

Rubenberg, Cheryl, 'Twenty years of Israeli economic policies in the West Bank and Gaza: prologue to the intifada', *Journal of Arab Affairs*, vol. 8, no 1 (1989), pp. 28–73.

Rubinstein, Danny, 'Protection racket, PA-style', *Ha'aretz* (3 November 1999).

Safieh, Afif, *The Peace Process: From Breakthrough to Breakdown* (London: PLO, 1997).

Safran, Nadav, *From War to War: The Arab–Israeli Confrontation 1948–1967* (New York: Pegasus, 1969).

Sahliyeh, Emile, *In Search of Leadership: West Bank Politics since 1967* (Washington DC: Brookings Institution, 1988).

Said, Edward, *Covering Islam: How the Media and the Experts Determine How We See the Rest of the World* (London: Routledge and Kegan Paul, 1981).

Said, Edward, *The End of the Peace Process: Oslo and After* (London: Granta, 2000).

Said, Edward, *Orientalism: Western Conceptions of the Orient* (London: Penguin, 1978).

Said, Edward, *Peace and its Discontents: Gaza–Jericho 1993–1995* (London: Vintage, 1995).

Said, Edward, *The Politics of Dispossession: The Struggle for Palestinian Self-Determination 1969–1994* (London: Vintage, 1995).

Said, Edward, *Power, Politics and Culture: Interviews with Edward Said*, ed. Gauri Viswanathan (New York: Pantheon, 2001).

Samara, Adel, 'Globalization, the Palestinian economy and the peace process', *Journal of Palestine Studies*, vol. 29, no 2 (2000), pp. 20–34.

Samara, Adel, *The Political Economy of the West Bank 1967–1982* (London: Khamsin, 1989).

Save the Children Fund West Bank and Gaza, 'Project Submission to DfID for Water and Sanitation Program in the Dura Area, Southern West Bank' (August 1998).

Savir, Uri, *The Process: 1,100 Days That Changed the Middle East* (New York: Random House, 1998).

Sayigh, Yezid, 'Armed struggle and state formation', *Journal of Palestine Studies*, vol. 26, no 4 (1997), pp. 17–32.

Sayigh, Yezid, 'The Palestinian economy under occupation: dependency and pauperization', *Journal of Palestine Studies*, vol. 15, no 4 (1986), pp. 46–67.

Scarpa, David, 'The southern West Bank aquifer: exploitation and sustainability', in J. Ginal and J. Ragep (eds), *Water in the Jordan Valley: Technical Solutions and Regional Cooperation* (Oklahoma: University of Oklahoma Press, 2002).

Scheumann, Waltina, and Manuel Schiffler (eds), *Water in the Middle East: Potential for Conflicts and Prospects for Cooperation* (Berlin: Springer, 1998).

Schiff, Ze'ev, *Security for Peace: Israel's Minimal Security Requirements in Negotiations with the Palestinians* (Washington DC: Washington Institute for Near East Policy, 1989).

Schiff, Ze'ev, 'Sharon suggests taking over water sources in West Bank', *Ha'aretz* (21 May 1997).

Schiff, Ze'ev, 'Water dispute deferred', *Ha'aretz* (8 September 1995).

Schiff, Ze'ev, and Ehud Ya'ari, *Intifada: The Palestinian Uprising – Israel's Third Front* (New York: Simon and Schuster, 1989).

Schwarz, Joshua, 'Israel Water Sector Study: Past Achievements, Current Problems and Future Options', *Report for the World Bank* (1990).

Schwarz, Joshua, 'Water resources in Judaea and Samaria and the Gaza Strip', in J. D. Elazer (ed.), *Judaea, Samaria and Gaza* (Washington DC: American Enterprise Institute, 1982).

Scobbie, Ian 'H_2O after Oslo II: legal aspects of water in the Occupied Territories', in *Palestine Yearbook of International Law* 1994/5, vol. 8 (Nicosia: Al-Shaybani Society of International Law, 1996), pp. 79–110.

Sen, Amartya, *Poverty and Famines: An Essay on Entitlement and Deprivation* (Oxford: Clarendon, 1981).

Serres, Michel, and Bruno Latour, *Conversations on Science, Culture and Time* (Ann Arbor: University of Michigan Press, 1995).

Sewell, James, *Functionalism and World Politics* (Princeton: Princeton University Press, 1966).

Sexton, R., 'The Middle East water crisis: is it the making of a new Middle East regional order?' *Capitalism, Nature, Socialism*, vol. 3, no 4 (1992), pp. 65–77.

Shafir, Gershon, *Land, Labour and the Origins of the Israeli–Palestinian Conflict, 1882–1914* (Cambridge: Cambridge University Press, 1989).

Shafir, Gershon, and Yoav Peled, *Being Israeli: The Dynamics of Multiple Citizenship* (Cambridge: Cambridge University Press, 2002).

Shafir, Gershon, and Yoav Peled, *The New Israel: Peacemaking and Liberalization* (Boulder CO: Westview, 2000).

Shalev, Michael, 'Have globalisation and liberalisation "normalised" Israel's political economy?', Israel Affairs, vol. 5, nos 2–3 (1998), pp. 121–55.

Shalev, Michael, 'The political economy of Labor Party dominance and decline in Israel', in T.J. Pempel (ed.), *Uncommon Democracies: The One-Party Dominant Regimes* (Ithaca: Cornell University Press, 1990), pp. 83–127.

Shapland, Greg, *Rivers of Discord: International Water Disputes in the Middle East* (London: Hirst and Co, 1997).

Shawwa, Isam, 'The water situation in the Gaza Strip', in Gershon Baskin (ed.), *Water: Conflict or Co-operation?* (Jerusalem: Israel/Palestine Center for Research and Information, 1992).

Shehadeh, Raja, *Occupier's Law: Israel and the West Bank* (Washington DC: Institute for Palestine Studies, 1988).

Sherman, Martin, *The Politics of Water in the Middle East: An Israeli Perspective on the Hydro-Political Aspects of the Conflict* (London: Macmillan, 1998).

Shikaki, Khalil, 'Old Guard, Young Guard: The Palestinian Authority and the Peace Process at Cross Roads', *PALDEV Digest*, no 2001-284 (1 November 2001).

Shiyoukhi, N., 'Palestinians protest water shortage', Associated Press (17 July 1999).

Shlaim, Avi, 'The Middle East: the origins of the Arab–Israeli wars', in Ngaire Woods (ed.), *Explaining International Relations Since 1945* (Oxford: Oxford University Press, 1996).

Shlaim, Avi, *The Politics of Partition: King Abdullah, The Zionists and Palestine, 1921–1951* (Oxford: Oxford University Press, 1990).

Shutz, Jim, *Globalization and the War for Water in Bolivia* (Cochabamba: Democracy Center, 2000).

Shuval, Hillel, 'Approaches to finding an equitable solution to the water resources problems shared by Israelis and the Palestinians in the use of the mountain aquifer', in Gershon Baskin (ed.), *Water: Conflict or Cooperation?* (Jerusalem: Israel/Palestine Centre for Research and Information, 1992), pp. 37–84.

Shuval, Hillel, 'Approaches to resolving the water conflicts between Israel and her neighbours – a regional water-for-peace plan', *Water International*, vol. 17, no 3 (1992), pp. 133–43.

Shuval, Hillel, 'Proposed principles and methodology for the equitable allocation of the water resources shared by the Israelis, Palestinians, Jordanians, Lebanese and Syrians', in Karen Assaf et al., *A Proposal for the Development of a Regional Water Master Plan* (Jerusalem: Israel/Palestine Centre for Research and Information, 1993), pp. 150–74.

Singer, J. David, 'The level of analysis problem in International Relations', in Klaus Knorr and Sidney Verba (eds), *The International System: Theoretical Essays* (Princeton NJ: Princeton University Press, 1961), pp. 77–92.

Smith, Charles, *Palestine and the Arab–Israeli Conflict*, 3rd edn. (New York: St Martin's Press, 1996).

Smith, Neil, *Uneven Development: Nature, Capital and the Production of Space*, 2nd edn. (Oxford: Blackwell, 1990).

Soffer, Arnon, 'The relevance of the Johnston Plan to the reality of 1993 and beyond', in Jad Isaac and Hillel Shuval (eds), *Water and Peace in the Middle East* (Amsterdam: Elsevier, 1994).

SOGREAH Ingenierie, 'Hebron Municipality Hydraulic Sketch' (1996).

SOGREAH Ingenierie, 'Master Plan for Water Distribution in the Bethlehem Area, Interim Report', vol. 1, Report for the PWA, Bethlehem 2000 Committee and the PWA (May 1998).

Sontag, Deborah, 'Quest for Mideast peace: how and why it failed', *New York Times* (26 July 2001).

Soper, Kate, *What is Nature?* (Oxford: Blackwell, 1995).

Sovich, Nina, 'Arafat's watch on Palestinian NGOs', *Palestine Report* (5 February 1999).

Starr, Joyce, *Covenant Over Middle East Waters: Key to World Survival* (New York: H. Holt, 1995).

Starr, Joyce, 'Water wars', *Foreign Policy*, no 82 (1991).

Starr, Joyce, and Daniel Stoll (eds), *The Politics of Scarcity: Water in the Middle East* (Boulder CO: Westview, 1988).

State Comptroller of Israel, 'Report on the Management of Water Resources in Israel' (Jerusalem: Government of Israel, 1990).

Stein, Kenneth, *The Land Question in Palestine, 1917–1939* (London: Chapel Hill, 1984).

Stork, Joe, 'Water and Israel's occupation strategy', *Middle East Report*, vol. 13, no 6 (1983), pp. 19–24.

Sullivan, Dennis, 'NGOs in Palestine: agents of development and foundation of civil society', *Journal of Palestine Studies*, vol. 25, no 3 (1996), pp. 93–100.

Swebmaoffer, Arnon, *Rivers of Fire: The Conflict over Water in the Middle East* (Lanham: Rowman and Littlefield, 1999).

Swyngedouw, Eric, 'Hybrid waters: on water, nature and society', presented at conference on Sustainability, Risk and Nature: The Political Ecology of Water in Advanced Societies, University of Oxford (15–17 April 1999).

Tamari, Salim, 'The revolt of the petite bourgeoisie: urban merchants and the Palestinian uprising', in Jamal Nassar and Roger Heacock (eds), *Intifada: Palestine at the Crossroads* (New York: Praeger, 1991), pp. 159–73.

Tamimi, Abdul Rahman, 'A Technical Framework for Final Status Negotiations over Water', *Palestine–Israel Journal of Politics, Economics and Culture*, vol. 3 (1996), pp. 70–2.

Todd, David, *Groundwater Hydrology* (New York: John Wiley and Sons, 1959).

Trottier, Julie, *Hydropolitics in the West Bank and Gaza Strip* (Jerusalem: PASSIA, 1999).

Trottier, Julie, 'Water and the challenge of Palestinian institution building', *Journal of Palestine Studies*, vol. 29, no 2 (2000), pp. 35–50.

Turner, Bryan, *Weber and Islam* (London: Routledge and Kegan Paul, 1974).

Turton, Anthony, 'Water Scarcity and Social Adaptive Capacity: Towards an Understanding of the Social Dynamics of Water Demand Management in Developing Countries', SOAS Water Issues Group, Occasional Paper No 9 (1999).

UNICEF, 'Water and Sanitation Briefing' (June 2001).

US National Academy of Sciences, *Water for the Future: The West Bank and Gaza Strip, Israel and Jordan* (Washington DC: National Academy Press, 1999).

USAID, 'Preliminary Findings of the Nutritional Assessment and Sentinel Surveillance System for the West Bank and Gaza' (5 August 2002).

USAID West Bank and Gaza Mission, 'USAID Contracts to Help Improve West Bank Water Systems', Press Release (17 July 1996).

USAID West Bank and Gaza Mission, 'USAID to Build New Water Supply Systems for the West Bank', Press Release (14 January 1998).

USAID West Bank and Gaza Mission, 'USAID-Funded Expansion of the Bethlehem-Hebron Water Supply System Complete', Press Release (5 December 1999).

USAID West Bank and Gaza Mission, 'Water Produced From the First Palestinian Owned and Operated Well in Bethlehem-Hebron Water Supply System', Press Release (26 July 1999).

Usher, Graham, 'Arafat's new cabinet – "back me or sack me"', *Middle East International* (21 August 1998), pp. 3–4.

Usher, Graham, *Dispatches from Palestine: The Rise and Fall of the Oslo Peace Process* (London: Pluto, 1999).

Usher, Graham, *Palestine in Crisis: The Struggle for Peace and Independence after Oslo*, 2nd edn. (London: Pluto, 1997).

Usher, Graham, 'Palestinian trade unions and the struggle for independence', *Middle East Report*, no 194/5 (1995), pp. 20–3.

Usher, Graham, 'The politics of internal security: the PA's new intelligence services', *Journal of Palestine Studies*, vol. 25, no 2 (1996), p. 23.

Usher, Graham, 'The rise of political Islam in the occupied territories', *Middle East International* (25 June 1993), pp. 19–20.

Usher, Graham, 'Running out of options?' *Middle East International* (1 September 2000).

Usher, Graham, 'Squeezing out the last drop', *Middle East International* (8 September 1995).

Vesilind, P., 'The Middle East's water: critical resource', *National Geographic*, vol. 183, no 5 (1993), pp. 38–71.

Vetter, Darci, 'The Impact of Article 40 of the Oslo II Agreement on Palestinian Water Provision' (Jerusalem: LAW, 1995).

Vidal, John, 'Water of strife', *The Guardian* (27 March 2002).

Wachtel, Boaz, 'The peace canal project: a multiple conflict resolution perspective for the Middle East', in Jad Isaac and Hillel Shuval (eds), *Water and Peace in the Middle East* (Amsterdam: Elsevier, 1994), pp. 363–73.

Wallerstein, Immanuel, *The Capitalist World Economy: Essays* (Cambridge: Cambridge University Press, 1979).

Wallerstein, Immanuel, *The Modern World System*, vols 1–3 (San Diego: Academic Press, 1974, 1980, 1989).

Waltz, Kenneth, *Man, The State and War* (New York: Columbia University Press, 1959).

Waltz, Kenneth, 'Realist thought and neorealist theory', *Journal of International Affairs*, vol. 44, no 1 (1990).

Waltz, Kenneth, *Theory of International Politics* (Reading MA: Addion-Wesley, 1979).

Weber, Max, *Economy and Society: A Outline of an Interpretive Sociology*, vol. 2 (New York: Bedminster, 1968).

Weber, Max, *The Methodology of the Social Sciences* (Glencoe: Free Press, 1949).

White House Press Release, 'Remarks by President Clinton et al' (13 September 1993).

Winner, Langdon, *The Whale and the Reactor: A Search for Limits in an Age of High Technology* (Chicago IL: Chicago University Press, 1986).

Winpenny, J.T., *Managing Water as an Economic Resource* (London: Routledge, 1994).

Wolf, Aaron, *Hydropolitics Along the Jordan River: Scarce Water and its Impact on the Arab–Israeli Conflict* (Tokyo: United Nations University Press, 1995).

Wolf, Aaron, and Masahiro Murakami, 'Techno-political decision making for water resource development: the Jordan River watershed', *Water Resources Development*, vol. 11, no 2 (1995), pp. 147–62.

World Bank, 'Developing the Occupied Territories: An Investment in Peace. vol. 1: Overview' (Washington DC: World Bank, 1993).

World Bank, 'Developing the Occupied Territories: An Investment in Peace. vol. 5: Infrastructure' (Washington DC: World Bank, 1993).

World Bank, 'Emergency Assistance Programme for the Occupied Territories' (Washington DC: World Bank, 1994).

World Bank, 'Fifteen Months: Intifada, Closures and Palestinian Economic Crisis – An Assessment' (March 2002).

World Bank, 'From Scarcity to Security: Averting a Water Crisis in the Middle East and North Africa' (Washington DC: World Bank, 1995).

World Bank, 'Strategic Water Resources Plan and Priority Investment Identification', Report for the PWA (1999).

World Bank, 'Strategic Water Resources Plan: Outline Terms of Reference, Draft' (16 March 1996).

World Commission on Environment and Development, *Our Common Future* (Oxford: Oxford University Press, 1987).

Worldwatch Institute, 'Populations Outrunning Water Supply', Press Release (23 September 1999).

WRAP, 'Hydrological Monitoring in Palestine: Status and Planning of the National Programme', report for the Palestinian Water Authority (June 1996).

WRAP, 'Palestinian Water Resources: A Rapid Interdisciplinary Sector Review and Issues Paper' (October 1994).

Young, Oran, *International Cooperation: Building Regimes for Natural Resources and the Environment* (Ithaca NY: Cornell University Press, 1989).

Zak, Moshe, 'Israeli–Jordanian negotiations', *Washington Quarterly*, vol. 8, no 1 (1985), pp. 167–76.

Zarour, Hisham, and Jad Isaac, 'Nature's apportionment and the open market: a promising solution to the Arab–Israeli water conflict', *Water International*, vol. 18, no 1 (1993), pp. 40–53.

Zarour, Hisham, and Jad Isaac, 'The water crisis in the Occupied Territories', Paper presented at the VII World Conference on Water, Rabat, Morocco (12–16 May 1991).

Zureik, Elia, *The Palestinians in Israel: A Study in Internal Colonialism* (London: Routledge and Kegan Paul, 1978).

Index